Studies in Social and Physical Anthropology

Studies in Social and Physical Anthropology

Essays in Memory of D.K. Sen

Editors

Gopala Sarana

R.P. Srivastava

RAWAT PUBLICATIONS

Jaipur • New Delhi • Bangalore • Hyderabad • Guwahati

ISBN 978-81-316-0448-9

Published by
Prem Rawat for **Rawat Publications**
Satyam Apts, Sector 3, Jawahar Nagar, Jaipur 302 004 (India)
Phone: 0141 265 1748 / 7006 Fax: 0141 265 1748
E-mail: info@rawatbooks.com
Website: www.rawatbooks.com

New Delhi Office
4858/24, Ansari Road, Daryaganj, New Delhi 110 002
Phone: 011 2326 3290

Also at *Bangalore, Hyderabad* and *Guwahati*

Typeset by Rawat Computers, Jaipur
Printed at Chaman Enterprises, New Delhi

Dilip Kumar Sen

(1921–1972)

Contents

Preface

This book has been brought out to commemorate the 90th birth anniversary of our teacher Dr D.K. Sen which fell on 30th May 2011. Sen took his Master's degree from Calcutta University and his Ph.D. from London University. He began his career as a teacher in the Department of Anthropology, Lucknow University, under the chairmanship of Dr D.N. Majumdar. After the sad demise of the latter, Sen chaired the Department for few years, and also edited the journal *The Eastern Anthropologist* before he was persuaded by his mentor N.K. Bose to join the Anthropological Survey of India. He soon became the Director of ASI and initiated some fruitful researches in Indian anthropology. He was elected President of Anthropology and Archaelogy Section of combined 51st and 52nd session of Indian Science Congress, Calcutta, 1964–65. He founded the Journal of Indian Anthropological Society and served as its editor until his untimely death in the year 1972 when he was only 51 years old. He died in harness as the cruel fate cut short a very promising career.

One may recall an instance narrated by P.K. Misra which shows Sen's gentle and protective attitude towards his junior colleagues and his readiness to step forth and fill a gap, if there is no one else to do so. In 1971, the silver jubilee of the Southern Regional Office of the ASI was celebrated at Mysore. Sen was then the Director of ASI and P.K. Misra was the Regional Officer. Dr V.K.R.V. Rao, the Central Minister for Education, inaugurated the programme. At the end of the meeting, it was discovered that no one had thought of arranging a lead singer for the national anthem. Dr Sen quickly grasped the situation. He got up from his seat and walked towards the microphone and began to sing the national anthem. By this time people sitting in the hall could seize the situation and joined Dr Sen in singing the national anthem. Dr Sen had a resonate and clear voice and the national anthem was very well rendered. Dr Sen not only saved the situation but also earned a lot of laurels from the media. The media asked him to give a press interview to which Dr Sen readily agreed. In the interview he outlined the work done by the Anthropological Survey of India, which was very well reported in the newspapers.

Wherever he worked, and in whatever capacity, he was loved by his students as well as colleagues, and consequently he could infuse good team spirit among them. Always cheerful, and with amiable ways and equable temperament, he was an *Ajatasatru* or a man without enemies. It was not possible to quarrel with him however much one might try. While he gave sound guidance to his students or junior researchers, he firmly refused to share authorship. He would say “it is your material. If I have given any guidance, you may acknowledge it, but the authorship is yours”, such was his professional ethic and scholarly integrity.

Some of his main research contributions are worth mentioning. He believed in population model and was opposed to typological approach. He argued that the skeletal material of Harappa era represented a homogeneous population and that it did not suggest any diversity of racial elements. His study of the blood group and haemoglobin variant of the Bengalees indicated their genetic relationship with other Indian populations. Sen was the first Indian anthropologist to decipher the role of random genetic drift with the help of demographic data from a village near Calcutta which has several endogamous castes. He found how the random genetic drift affected the gene frequencies of the two castes groups in this village. The implication of his findings is that we should not invoke the role of genetic drift without reliable data.

Dr Sen is survived by his wife Dr Tulika Sen an able anthropologist now a retired faculty of the Department of Anthropology, Calcutta University which was her *alma mater*. She has sustained him with unfailing devotion and encouragement. We wish her good health and a happy long life.

Acknowledgement

The editors invited articles for D.K. Sen festschrift from his colleagues and students as well as some anthropologists from abroad to contribute an article on a topic of their own choice. The response was very encouraging and the articles were received before the deadline. The editors acknowledge the contributions of the following authors with deep appreciation and express their gratitude. Senior Editor Prof. Gopala Sarana died on November 7, 2010 due to cardiac arrest. It is a great pity that he will not be with us to see this volume in print. It is hoped that the papers presented in this volume will be of interest to scholars in anthropology as well as cognate disciplines.

It goes without saying that the publisher Mr Pranit Rawat deserves our hearty thanks for readily taking up the publication.

Contributors

Ian Tattersall, Division of Anthropology, American Museum of Natural History, New York, NY 10024, USA.

Ian Gilligan, School of Archaeology and Anthropology, A.D. Hope Building 14, The Australian National University, Canberra, Australia.

Charles M. Musiba, Department of Anthropology, University of Colorado Denver, CO 80217–3364, USA.

Carol Laciny, Laboratory of Biological Anthropology, University of Colorado Denver, CO 80217–3364, USA.

Cassian C. Magori, Department of Anatomy and Histology, Weill-Bugando, University College of Health Sciences, Mwanza, Tanzania.

R.P. Srivastava, Former Professor of Anthropology, University of Poona, Pune.

Indera P. Singh, Former Professor of Anthropology, University of Delhi, New Delhi.

N. Subba Reddy, Former Professor of Anthropology, Madras University, Chennai.

Ravindra K. Jain, Former Professor of Anthropology, Jawaharlal Nehru University, New Delhi.

Gouranga P. Chattopadhyay, Former Professor of Anthropology, Indian Institute of Management, Kolkata.

P.K. Misra, Former Professor of Anthropology, North-Eastern Hill University, Shillong.

Part I

Physical Anthropology and Archaeology

1 Patterns in Human Evolution and the Great Leap Forward

Ian Tattersall

We human beings are very peculiar creatures. For one thing, we walk upright on two legs, a very eccentric thing for any member of the animal realm to do. We have great globular heads that balance precariously atop our curiously S-shaped vertical spines, our females conceal their ovulation, we clothe and decorate ourselves, and we make elaborate tools to accomplish tasks that other creatures use their beaks or teeth for. The list goes on and on. But most bizarre of all is the way we process information in our brains. All other creatures we know of on Earth, even such cognitively sophisticated ones as chimpanzees, live in the world essentially as Nature presents it to them. They may react to sensory stimuli in highly sophisticated ways, but they accept the world as it is. Not us. Uniquely, as far as we know, we *Homo sapiens* dissect all that our perceptions of the world in our minds, creating mental symbols that we can then combine and recombine to create new realities or alternate constructions of the one we are in. In a very literal sense, we human beings – and we alone – live in worlds that are at least partially of our own making. And none of us interprets what surrounds us in exactly the same way.

There is, then, a great cognitive gulf between us and all the other denizens of the planet – even our closest living relatives, the great apes. We are symbolic creatures, they are not. Yet only those with an axe to grind in this matter would contest that we are descended, at some remove, from an ancestor that did not share our cognitive peculiarities – an ancestor that lay on the other side of that narrow but infinitely deep gulf. So how was the chasm crossed? Well, as long as we do not know how a mass of electrochemical signals in our brains becomes transformed into what we each individually experience as our subjective consciousness, we will never be able to answer that question mechanistically. But at this point, we do know enough about our biological past – about those now-extinct relatives who preceded us on Earth – to hazard at least a framework, and even a timescale, within which that fateful passage was made.

In the beginning

Let us first consider the starting point. The cognitive scientist Daniel Povinelli recently proposed that a fundamental distinction between the ways in which chimpanzees and humans view the world is that, while humans form abstract views about other individuals and their motivations, "chimpanzees rely strictly upon observable features of others to forge their social concepts. If correct, [this] would mean that chimpanzees do not realize that there is more to others than their movements, facial expressions, and habits of behaviour. They would not understand that other beings are repositories of private, internal experience" (Povinelli, 2004: 33). It would also imply that individual chimpanzees do not have such awareness of themselves either. They *experience* the emotions and intuitions that arise in their own minds; and they may act on them, or suppress them, as the situation demands or permits. But they "do not reason about what others think, believe and feel – precisely because they do not form such concepts in the first place" (Povinelli, 2004: 33).

This is about as much as we can reasonably say about chimpanzees' minds from our outside vantage point. Still, even though there has doubtless been a lot of evolutionary water under the bridge on both sides since human beings shared an ancestor with any ape, most authorities find it reasonable to conclude that cognition of the kind we see among chimpanzees provides us with a reasonable approximation of the cognitive point from which our ancestors started some seven million years ago. To use Povinelli's words (2004: 34), one may reasonably assume that those ancestors were "intelligent, thinking creatures who deftly attend[ed] to and learn[ed] about the regularities that unfold[ed] in the world around them. But … they [did] not reason about unobservable things: they [had] no ideas about the 'mind,' no notion of 'causation'." In the human sense, they had, as yet, no idea of self. This plausible yet necessarily sketchy characterization of our lineage's starting point more or less exhausts what can usefully be said on the basis of existing studies of comparative cognition. If we wish to know how we got here from there, we need to turn to our fossil and archaeological records.

The current best estimate is that our most recent common ancestor with any ape lived in the order of seven to eight million years ago. In the first half of this period, a handful of mostly poorly known African species contend to be the "earliest" hominid, among them the recently ballyhooed and very peculiar *Ardipithecus ramidus* (White et al., 2009). Now known from one crushed skeleton and the remains of several other individuals, all some 4.4 million years old, this primate seems to have lived in a fairly densely forested environment and is estimated to have weighed in at over 50 kg. And it defeats all expectations about the adaptation of any very early hominid. Most palaeoanthropologists reckon that the closest living relatives of human beings are the chimpanzees; and if this is so, as they go back in time, fossil humans ought to be more chimpanzee-like, at least in certain limited respects. Yet, *Ardipithecus* is reported to be decidedly un-chimpanzee-like in a whole host of features. Its hand, for instance, is said to show none of the specializations for knuckle-walking exhibited by the hands of African apes – though at the same time it is reported not to show any suspensory features in its forelimb, something very curious indeed for an arboreal primate of large body size. Evidently, the common ancestor must have boasted a much more generalized structure than any living ape.

Also, the *Ardipithecus* does not much resemble the archetypal early hominid *Australopithecus afarensis*, best exemplified by the 3.2 million-year-old fossil "Lucy," the previous holder of the "earliest hominid skeleton" title. Lucy, while retaining some features that indicate her descent from a tree-living ancestor, has a pelvis and leg bones that clearly indicate that she walked upright when on the ground; and the same is true of foot bones belonging to other individuals of her species. This is not the case with *Ardipithecus* whose foot shows a clearly divergent (grasping) great toe rather than the in-line configuration typical of hominids and whose pelvis, allowing for the vagaries of reconstruction, appears to have been as much like that of an ape as like Lucy's. Moreover, its one claimed definitively bipedal feature (the presence of a distinct anterior inferior iliac spine) is also present in the pelvis of the Miocene possible nonhominoid *Oreopithecus* (which also has a widely splayed great toe). Complicating matters yet further is the fact that the Kenyan fossil *Australopithecus anamensis*, not much younger than *Ardipithecus* at 4.2 million years old, appears in known parts (basically, jaws and leg bones) to be very much more like Lucy, to whose species it is plausibly ancestral. If this is borne out by further finds, *Ardipithecus* will be clearly seen to represent a weird side-branch of early hominid evolution.

In the basic structure of its skull, though not in many details, *Ardipithecus* is claimed to resemble the six-to-seven-million-year-old *Sahelanthropus* from Chad, in central-west Africa, which shocked Palaeoanthropology seven years ago by combining a tiny braincase with an apparently short face such as might have been expected in a hominid several million years younger [but since elongated somewhat in a "virtual" reconstruction (Zollikofer et al., 2005)]. At six million years old *Orrorin*, a poorly known form from Kenya, is also dentally different and possesses the earliest leg bones that plausibly indicate uprightness. If the *Ardipithecus* lineage goes back as far as claimed, we thus have three contenders for very early hominid status at around six million years ago, all of them very different from each other and from the first *Australopithecus*. And clearly, the early hominid story was not that of a single lineage becoming steadily better adapted to life on the ground with passage of time.

All in all, then, the extraordinary *Ardipithecus ramidus* emphasizes quite dramatically that, from the very beginning, the evolutionary history of the hominid family has been one of vigorous evolutionary experimentation with an evidently highly malleable *Bauplan*. This is a pattern that has energetically carried through all the way to the origin of *Homo sapiens*, a species that is quite as extraordinary in its own way (Figure 1). Taken together, the earliest hominids make an oddly assorted group, with the clear implication that, from the very beginning, the history of the human family has actually been fairly standard for a mammalian group that has thrived as the hominids had done. Just as most successful mammalian families are rather diverse, right up until very recently the Earth has routinely hosted a variety of different hominids, each battling in its own way for ecological space. Of course, many of us were brought up with the idea that (most unusually) human evolution involved the gradual modification of a single lineage that culminated in today's lonely *Homo sapiens*; but this notion is flatly contradicted by the facts of a rapidly expanding fossil record (Figure 1). Indeed, as recently as 30 thousand years ago there were at least four hominid species on Earth; and our present solitary state, I am convinced, tells us something very significant about ourselves, about just how *different* we are from *any* of our precursors.

The first really well-documented hominid species is *Australopithecus afarensis*, whose fossils span a time range of about 3.6 to 3 million years ago. Emblematic of a whole host of species that flourished in the African woodlands between about 4 and 1.5 million years ago, members of *A. afarensis* were diminutive creatures that would have been adept climbers on trees even while clearly walking upright on the ground, albeit not exactly as we do today. Early hominids of this general kind are often informally known as "bipedal apes" because whatever the modifications of their hips, legs, and feet, their cranial proportions were ape-like, with small braincases and large, protruding faces. In terms of cognition, there is no reason to suspect that these creatures were significantly smarter than any modern ape.

Figure 1

Very tentative family tree of hominids, showing the variety of extinct species now known within the human family and their distribution in time, and sketching in some possible lines of descent.

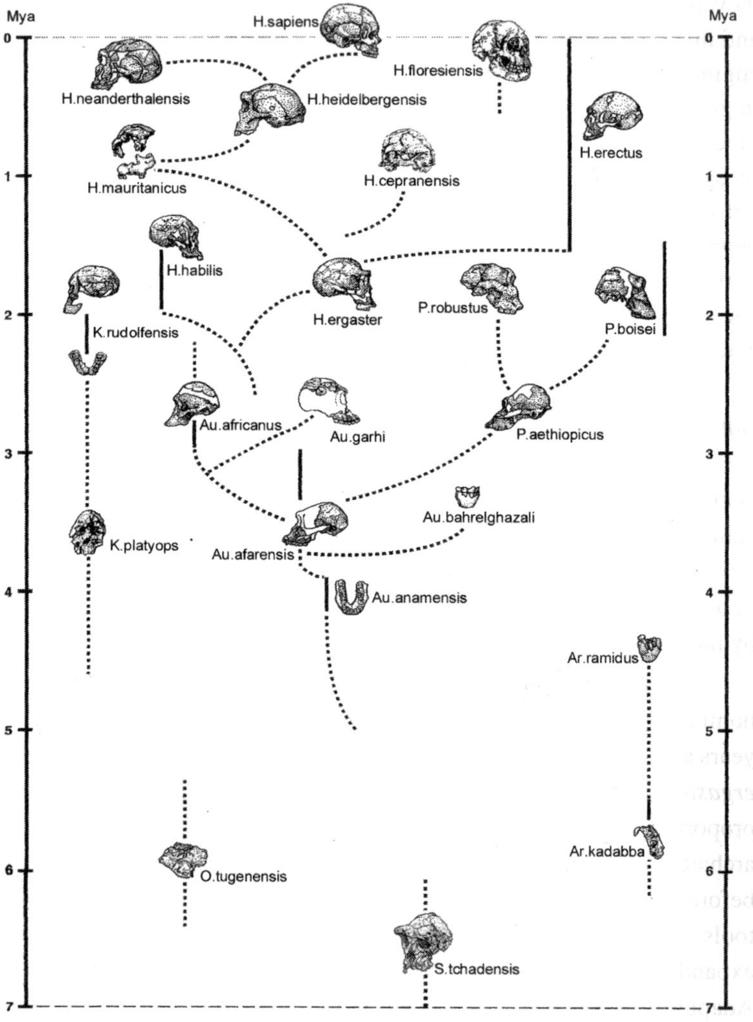

The first stone toolmakers

Yet, at some point something happened. At around 2.5 million years ago, early African hominids started making stone tools. No doubt that wooden and other tools were already used by hominids long before, as they are used by apes today; but those early stone tools preserved well, and marked a major (and fateful) step ahead in hominid behaviour and cognition. Even with intensive coaching, no living ape has been able to grasp the idea of hitting one carefully selected rock with another, at precisely the angle required to detach a sharp flake. What is more, those carefully selected stones were often carried long distances before being made into tools as required, implying an impressive degree of foresight on the part of the toolmakers. Without any doubt, with the first stone tools we have the earliest good evidence that our ancestors had advanced cognitively far beyond the ape league (Schick and Toth, 1993).

We do not know exactly who those first stone toolmakers were. But there is good circumstantial evidence that they were creatures who we would almost certainly classify physically as "bipedal apes" (Asfaw et al., 1999). Which means that, yet again, we have early evidence of a pattern that would hold throughout the rest of human evolution: the arrival of new ways of doing things is never explained by the evolution of a new kind of hominid. Technological innovations are always made by members of pre-existing species.

So, it should come as no surprise that when a totally new kind of hominid came on the African scene at some time after about two million years ago, it did not bring with it a new tool kit. At the beginning, *Homo ergaster*, the first hominid species to have effectively modern body proportions, made simple flake tools more or less exactly like those its archaically bodied predecessors had been making for a million years before. Establishing a routine practice for hominids, it adapted the old tools to new ways, which would have included a life out on the expanding African savannas, away from the shelter of the trees. Adapted like us not merely to uprightness, but to a striding locomotion over open ground, these hominids would have had a lifestyle that emphasized stamina and endurance, a hallmark of hominids ever since.

Out of Africa

Once the modern body type was established, hominids spread rapidly beyond Africa: they were in the Caucasus by about 1.8 million years ago (de Lumley et al., 2002) and in Indonesia not much later (Swisher et al., 1994). By 1.4 million years ago, they had already penetrated the harsher climes of Europe (Carbonell et al., 2008). Interestingly, this initial spread was achieved in the absence of radically larger brains or improved technologies. Still, back in Africa, and a bit later elsewhere, we do see the inauguration of a trend among hominids towards increasing brain size. This trend seems to have been expressed independently in several hominid lineages in different parts of the world, so that there is evidently something about the genus *Homo* that has historically predisposed it to increasing brain sizes. Brain is metabolically an expensive tissue that takes a lot of energy to maintain, and it will be very important to understand just what it is that underwrote this trend if we are ever going to fully unravel the story of the evolution of human cognition. But right now what that underpinning element was remains unclear, although it must almost certainly have had in some way to do with a feedback between human lifeways and a form or forms of "intelligence" that we cannot yet specify (Tattersall, 2008).

Again in Africa, a radically new type of tool was invented about 1.5 million years ago, still during the tenure of *Homo ergaster*. This was the "handaxe," a largish tool made by carefully shaping a stone "core" on both sides to flat, teardrop shape (Klein, 1999). The earliest stone tools were no more than simple flakes with a sharp cutting edge, the only feature desired. It did not matter what shape the tool was. But handaxe-makers were fashioning tools to a specific pattern that existed in the toolmaker's mind before production started. This was clearly a conceptual advance, but alas we do not know to what extent it may have reflected more complex cognitive processes on the part of the hominids concerned.

The same can be said of the next major conceptual advance in stone toolmaking, the "prepared-core" tool, whereby a stone core was carefully shaped on both sides until a single blow would detach a more or less finished tool with a cutting surface all around its periphery. Invented at some time after about 300 thousand years ago, following a long period of stasis in stone-tool production, prepared-core tools were introduced during the tenure of *Homo heidelbergensis* (Klein, 1999). This, the first cosmopolitan hominid species, appeared in Africa about 600 thousand years ago, and rapidly spread to Europe and eventually as far afield as China (Tattersall and Schwartz, 2009).

Members of *Homo heidelbergensis* had brains that lay within the lower limit of the modern size range, but at the beginning they were associated with archaic tool types in Africa, and in Europe they started making handaxes fairly late. The earliest evidence for the sustained control of fire in hearths goes back to about 800 thousand years ago (Goren-Inbar et al., 2004), before we have any evidence of *Homo heidelbergensis*, but the routine use of fire in this way dates only to about 400 thousand years ago, well within the tenure of this species. Also of about this age are the first constructed shelters (de Lumley and Boone, 1976) and the first-known wooden throwing spears (Thieme, 1997), both equally significant technological innovations. Clearly, something was astir cognitively at around this time; but again we are frustratingly unable to tie it in to the way *Homo heidelbergensis* actually experienced the world around it. Still, it is almost certainly significant that no symbolic artefacts are associated with *Homo heidelbergensis*.

Quite probably, the most sophisticated practitioners ever of the prepared-core technique were the Neanderthals, *Homo neanderthalensis*, who flourished in Europe and western Asia between about 200 thousand and 30 thousand years ago. These hominids had brains as large as ours, albeit housed in skulls of rather different aspect and differing from us in various aspects of body structure (see Figure 2). At least occasionally, and very simply, they buried their dead; and there is evidence that they looked after disadvantaged members of the social group. They were excellent craftsmen, though apparently rather unimaginative – their beautiful products always look remarkably similar, wherever they come from. It was these hominids who were apparently abruptly displaced by the first *Homo sapiens* to enter Europe, about 40 thousand years ago – whether by direct conflict or by indirect economic competition is not known.

Figure 2

Comparison of a reconstructed Neanderthal skeleton (left) and that of a modern human of similar stature. Note the substantial differences in structure, particularly in the pelvic and thoracic regions. In most features, the Neanderthal probably represents the primitive condition for species in the genus Homo; it is the Homo sapiens skeleton that is highly derived.

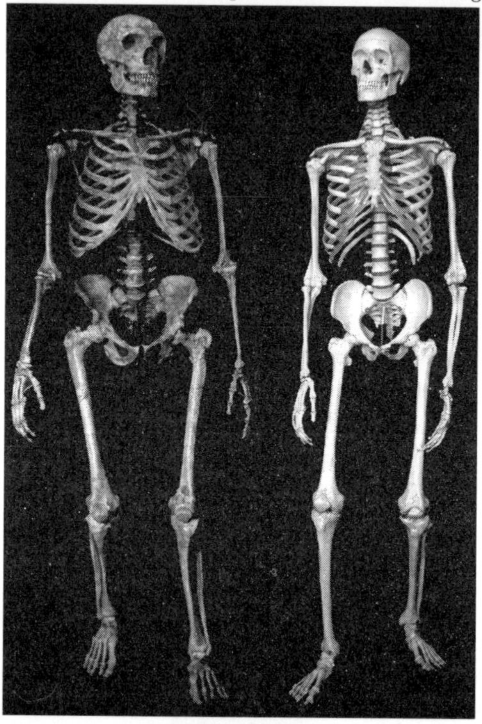

Courtesy: AMNH

The dual origin of modern *Homo sapiens*

The contrast between the Neanderthals and the invading "Cro-Magnons" was extreme, principally because the latter had lives that were quite overtly pervaded by symbol. By 35 thousand years ago, they were carving exquisite figurines and painting powerful animal images on cave walls; soon they were decorating everyday objects, making elegant pictorial and geometric engravings, and developing systems of notation. On the economic front, they were occupying the landscape in far greater density than their predecessors, and exploiting its resources with greater efficiency and perception. In short, the Cro-Magnons were *us*, while the nonsymbolic Neanderthals remained emblematic of all the other hominids that had preceded them: clever for their times, certainly, but lacking the creative spark that makes us who we are (Tattersall, 2008).

That spark was not acquired by *Homo sapiens* in Europe. Instead it was brought in with them by the invaders from a place of origin that almost certainly lay in Africa where our species itself evolved. Fossils from Ethiopia show that, as an anatomically recognizable entity, *Homo sapiens* had emerged by about 200–160 thousand years ago (White et al., 2003; McDougall et al., 2005). But those early *Homo sapiens* are not associated with archaeological remnants that suggest that they were symbolic. Instead, they did as their predecessors had done and adapted to changing environmental conditions by using traditional technologies in new ways. This is also true of the earliest *Homo sapiens* who penetrated beyond Africa into the Levant at the eastern end of the Mediterranean. There the newcomers, documented by about 93 thousand years ago (Valladas et al., 1988), did not immediately displace the resident Neanderthals, who persisted for another 50 thousand years; and they also made virtually identical stone tool kits. There is no reason to suspect any cognitive difference between *Homo sapiens* and *Homo neanderthalensis* at that time and place.

The first stirrings of symbolic behaviour are found in southern Africa; and although not associated with human fossils, they are almost certainly attributable to early *Homo sapiens*. Overtly, symbolic objects include a couple of 77 thousand-year-old ochre plaques engraved with geometric motifs from Blombos cave on the southern African coast (Henshilwood et al., 2002) and shells pierced for stringing (hence for bodily decoration) come from that (Henshilwood et al., 2004) and other sites of similar age. Complex heat treatment of silcrete from the nearby Pinnacle Point, dated to more than 70 thousand years ago (Brown et al., 2009), also seems to indicate very complex sequential thought processes at this time.

Observations like these suggest that full modern humanity was achieved in two stages. First, the very distinctive modern human anatomy was acquired, plausibly as a result of a change in gene expression that had cascading developmental consequences throughout the body. If that change, minor in structural genetic terms, involved neural circuits as well as the structures visible in the fossilized skeleton, it would have created a new cognitive potential. Evidently, though that potential was not immediately expressed. This would be nothing unusual: changes involving cognition in hominid history had, after all, always been backwardly compatible: new hominids had routinely continued with the technological habits of their predecessors before discovering new ways of doing things. Still, the new potential, nonetheless, had to be *discovered* by its possessor; and that had to have happened through a *cultural* stimulus (the biology was, after all, already there). It was this act of behavioural discovery that announced the second stage of becoming fully human, some tens of thousands of years after anatomical modernity had been achieved (Tattersall, 2008). My favourite candidate for the stimulus involved is the invention of language – which, depending as it does on the creation of mental symbols for its very existence, is the ultimate symbolic activity. What is more, language is a supremely communal attribute that, once adopted, would have very easily spread through a population with the biological predisposition to absorb it.

Emergence in the origin of human cognition

The various factors just considered put the origin of the extraordinary and radically unique human cognitive capacity in the context of emergence, whereby something entirely new is created from a novel combination of elements – much as the final placement of the keystone of an arch creates a structure with a radically new function. After some 400 million years of accretionary vertebrate, mammal and primate brain evolution, the human brain had evolved to a point where a small addition was able (somewhat like the keystone of an arch) to create a structure with an entirely novel potential: in this case, for a radically new way of processing information about the world and about its own internal state. Many commentators, including Alfred Russel Wallace, co-inventor of the idea of evolution by natural selection, have had difficulty imagining how selection could have driven human consciousness into existence. The substitution of emergence for selection now places that fateful event in an entirely routine evolutionary context: one that requires no special explanation. And it focuses scientific attention where it ought to be: on the precise identity of the alteration to the brain that enabled human beings to become the extraordinary creatures they are.

Acknowledgements

I thank Dr Rajeshwar Srivastava for this opportunity to help celebrate the memory and distinguished career of Dr D.K. Sen. Parts of this essay have been abstracted for a general audience in *The Contemporary West*.

References

Asfaw, B., White, T., Lovejoy, O., Latimer, B., Simpson, S., and Suwa, G. 1999. *Australopithecus garhi*: A new species of early hominid from Ethiopia. *Science* 284: 629–635.

Brown, K.S., Marean, C.W., Herries, A.I.R., Jacobs, Z., Tribolo, C., Braun, D., Roberts, D.L., Meter, M.C., and Bernatchez, J. 2009 Fire as an engineering tool of early modern humans. *Science* 325: 859–862.

Carbonell, E., et al. 2008. The first hominin species of Europe. *Nature* 452: 465–470.

de Lumley, H., and Boone, Y. 1976. Les structures d'habitat au Paléolithique inférieur. In H. de Lumley (ed.) *La Préhistoire frančaise* (vol. 1). pp. 635–643. Paris: CNRS.

de Lumley, H., Lordkipanidze, D., Féraud, G., Garcia, T., Perrenoud, C., Falguères, C., Gagnepain, J., Saos, T., and Voinchet, P. 2002. Datation par la méthode 40^Ar/39^Ar de la couche de cendres volcaniques (couche VI) de Dmanissi (Géorgie) qui a livré des restes d'hominidés fossils de 1.81 Ma. *C.R. Palevol* 1: 181–189.

Goren-Inbar, N., Alperson, N., Kislev, M.E., Simchoni, O., Melamed, Y., Ben-Nun, A., and Werker, E. 2004. Evidence of hominin control of fire at Gesher Benet Ya'aqov, Israel. *Science* 304: 725–727.

Henshilwood, C.S., d'Errico, F., Yates, R., Jacobs, Z., Tribolo, C., Duller, G.A.T., Mercier, N., Sealy, J.C., Valladas, H., Watts, I., and Wintle, A.G. 2002. Emergence of modern human behaviour: Middle Stone Age engravings from South Africa. *Science* 295: 1278–1280.

Henshilwood, C., d'Errico, F., Vanhaeren, M., van Niekerk, K., and Jacobs, Z. 2004. Middle Stone Age shell beads from South Africa. *Science* 304: 404.

Klein, R.G. 1999. *The Human Career*, 2nd ed. Chicago, IL: University of Chicago Press.

McDougall, I., Brown, F.H., and Fleagle, J.G. 2005. Stratigraphic placement and age of modern humans from Kibish, Ethiopia. *Nature* 433: 733–736.

Povinelli, D.J. 2004. Behind the ape's appearance: escaping anthropocentrism in the study of other minds. *Daedalus* 133(1): 29–41.

Schick, K.D. and Toth, N. 1993. *Making Silent Stones Speak: Human Evolution and the Dawn of Technology*. New York: Simon and Schuster.

Swisher C.C., Curtis, G.H., Jacob, T., Getty, A.G., Suprijo, A., and Widiasmoro. 1994. Age of the earliest known hominids in Java, Indonesia. *Science* 263: 1118–1121.

Tattersall, I. 2008. An evolutionary framework for the acquisition of symbolic cognition by *Homo sapiens. Comparative Cognition & Behaviour Review* 3: 99–114.

Tattersall, I. and Schwartz, J.H. 2009. Evolution of the genus *Homo. Annual Reviews of Earth and Planetary Sciences* 37: 67–92.

Thieme, H. 1997. Lower Palaeolithic hunting spears from Germany. *Nature* 385: 807–810.

Valladas, H., Reyss, J.L., Joron, J.L., Valladas, G., Bar-Yosef, O., and Vandermeersch, B. 1988. Thermoluminescence dating of Mousterian "Proto-Cro-Magnon" remains from Israel and the origin of modern man. *Nature* 331: 614–616

White, T.D., Asfaw, B., DeGusta, D., Gilbert, H., Richards, G.D., Suwa, G., and Howell, F.C. 2003. Pleistocene *Homo sapiens* from Middle Awash, Ethiopia. *Nature* 423: 742–747.

White, T.D. et al. 2009. Special Issue on *Ardipithecus ramidus. Science* 326: 5–106.

Zollikofer, C.P.E., Ponce de León, M.S., Lieberman, D.E., Guy, F., Pilbeam, D., Likius, A., Mackaye, H.T., Vignaud, P., and Brunet, M. 2005. Virtual cranial reconstruction of *Sahelanthropus tchadensis*. *Nature* 434: 755–759.

2 Climatic Variation in Femoral Neck-Shaft Angle Among Modern Humans

Continental and Regional Trends

Ian Gilligan

The role of environmental adaptation as an influence on human morphology has long been of central interest in physical anthropology, and the possible effects of climatic variables such as changing temperature regimes (past and present) remain a major focus of research. A key aspect of morphological adaptation to the thermal environment involves the relationship between body shape and heat loss, with a more linear body form being more efficient in losing excess heat to the surroundings. Conversely, a comparatively rounded or stocky shape exposes less surface area for a given volume or mass, which is biologically advantageous in circumstances where heat conservation is a priority. At a population level, modern human groups can be differentiated according to overall variation in various measures of body shape.

The author is an MBBS too.

To some extent, these broad differences can be seen as reflecting long-term morphological patterning in relation to geographical distributions (assuming migration histories are taken into account) and, ultimately, to physiologically relevant climatic variables, notably temperature indices.

Numerous studies in physical anthropology have explored the possible adaptive relationships between morphology and climate by comparing body shape measures in different human groups and relating these to climatic indices or proxies such as latitude and average annual temperatures. The degree to which certain measures of body shape might be useful for this purpose is a matter of some debate, as climatic adaptation is only one of a number of factors which are known to exert an influence on body form. The others include, for instance, patterns of nutrition, diet, disease, lifestyle, and levels (and types) of physical activity, particularly during early development and childhood. With respect to skeletal measures of body shape, one poorly-researched measure that has recently attracted considerable attention in the literature concerns the shape of the femur and, specifically, the angle formed between the neck and the shaft at its proximal end – designated variously over the years as the collodiaphyseal angle, cervicodiaphyseal angle, or the angle of inclination, and nowadays known widely as the neck-shaft angle (NSA).

This chapter attempts to establish the nature and extent of variation in the NSA between and within modern human groups and, insofar as such variation can be demonstrated reliably with available data, examine whether the evidence might support the utility of the NSA as an osteological indicator of morphological adaptation to thermal conditions. Further questions relating to the possible confounding effect of behavioural adaptations (e.g., the use of clothing) on any relationship between morphological and environmental variation will also be discussed briefly. However, the present report is limited essentially to the presentation of primary data obtained from an extensive global survey of the NSA among modern human groups, and the discussion of results from an exploratory statistical analysis of the findings in relation to simple climatic indices (latitude, mean annual temperature, and mean summer and winter temperatures).

Background

The adaptive relationship between body shape and climate is known in biology as "Bergmann's rule", based on observations of the importance of the surface area to volume ratio in determining an organism's heat balance (Bergmann, 1847: 601). Similarly, Allen's rule refers to the size of body appendages, with exposed limbs becoming shorter in cooler climates (Allen, 1877). The historical background and physiological principles are summarized in Gould (1966) and Ruff (1994). Briefly, body form affects thermal requirements by virtue of the ratio between skin surface area and body volume, or mass. The latter is related to total heat production (through metabolism), while exposed skin surface area plays a dominant role in dissipating body heat. Even minor variations in body shape or size are significant because, as Bergmann observed, surface area and volume vary in unequal proportions to each other: volume varies as the cube, whereas surface area varies as the square, of any linear change in size. These principles predict a lighter, more linear body build in warmer environments, and a heavier and more stocky build in cooler regions.

Like all rules that of Bergmann has exceptions, as emphasized by critics (e.g., Geist, 1987). With increasing latitude, for instance, body mass first increases but then decreases in some species. This reflects the fact that greater body mass attracts a higher caloric cost, so its net adaptive benefits reach limits, at which point alternative strategies such as increased fur cover can be favoured. Phenotype is "the result of a compromise between many conflicting selection pressures" (Mayr, 1956: 106), and exceptions serve only to illustrate the complexities involved (e.g., Schreider, 1975; Katzmarzyk and Leonard, 1998). With this caveat, Bergmann's rule has been shown to apply among numerous bird and mammal species (e.g., Ashton et al., 2000; Meiri and Dayan, 2003), while studies among modern humans have demonstrated that body mass and shape, head shape, and relative limb proportions correlate with thermal conditions (e.g., Hiernaux, 1963; Roberts, 1978; Crognier, 1981; Houghton, 1990; Ruff, 1994, 2002). Climate-related trends in body shape (Figure 1) have been shown to exist not only between but also within modern human groups, with the Australian Aborigines being a prime example. The long time depth of their

Figure 1

Differing body shapes and limb proportions among modern human groups, with the ratios of surface area to body mass (cm^2/kg) (redrawn from Ruff, 1993: 54)

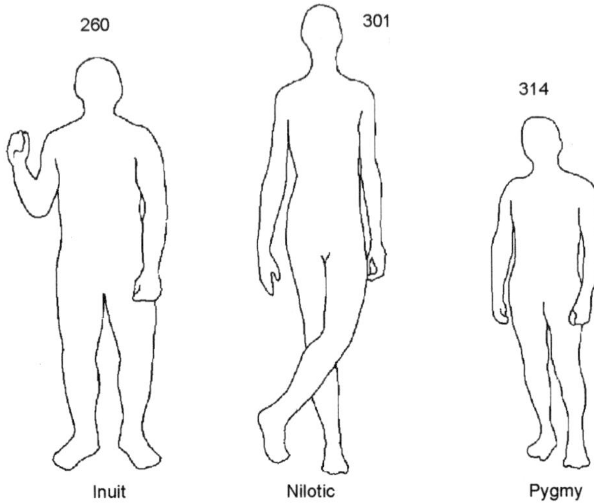

presence on the continent – at least 45,000 years (O'Connell and Allen, 2004) – has allowed adequate time for the emergence of a wide range of morphological trends which correlate with climatic indices (Gilligan and Bulbeck, 2007). Limb proportions have also been shown to vary with altitude, for instance between coastal and highland groups in Peru (Weinstein, 2005).

Late Pleistocene Hominins

Among fossil hominids, indices of body shape and limb proportions have been shown to correlate with thermal conditions (e.g., Ruff, 1991, 1994; Holliday and Falsetti, 1995). For example, the brachial and crural indices of "cold adapted" Neanderthals are in the lower range compared with both early and contemporary *Homo sapiens*, and the indices are lower among European than Near Eastern Neanderthals (Trinkaus, 1981: 189–209). A similar trend is evident in the claviculo-humeral index (ibid.). For the latter, a high index is an indicator of a more stocky body build. It is greater in Neanderthals compared with modern humans in late Pleistocene Europe, and also in Europeans compared with Near Eastern Neanderthals. Among earlier hominins, the tall, linear build of *H. ergaster* was well-adapted to open, semi-arid environments, whereas the shorter stature and the limb proportions of the Australopithecines were adapted to more closed, wet-tropical environments (Ruff, 1993: 56–58). With regards to head size and shape, a trend to larger and rounder heads is a general feature of hominid evolution associated with brain expansion and the adoption of an erect posture (Weidenreich, 1945), with a further likely effect of cold adaptation among more recent hominids throughout the Pleistocene (Beals et al., 1983).

Fully modern humans who replaced Neanderthals in late Pleistocene Europe retained tropical limb proportions – e.g., a high crural index (tibial length × 100 × femoral length) – but nonetheless showed a small and gradual reduction in total lower limb length, which may reflect "climatic selection due to the glacial cold" (Holliday, 1999: 561). Indeed, the Neanderthal-like limb proportions of the juvenile skeleton dated to around 25,000 years ago in Portugal (Lagar Velho 1, known as the "Lapedo child") were among the unusual features that first alerted its discoverers to its possible status as "hybrid" between cold-adapted Neanderthals and morphologically tropical modern humans (Ruff et al., 2002: 390; Zilhão and Trinkaus, 2002: 21), for which further anatomical evidence has since emerged (e.g., Bayle et al., 2010).

Body shape and clothing

Adaptation to thermal environments can be behavioural as well as biological, and both these basic evolutionary strategies can interact. Since at least the middle Pleistocene and particularly the late Pleistocene, humans have developed increasingly elaborate (and thermally effective) forms of artificial insulation from the environment, concomitant with a generally increasing pace of social and technological change, notably in recent millennia. One likely consequence is the reduction in selection pressure for biological adaptation to external environmental conditions. This process has allowed the expansion of hominins from their tropical and subtropical homelands into temperate and subpolar regions, with some populations eventually penetrating and occupying high latitudes even during colder phases within the last couple of glacial cycles. In the modern world, the success of behavioural measures resulting in greater protection from (and modification of) local thermal conditions should become visible as a trend towards a decoupling of morphological and climatic variation, and this should be most discernible among those human groups utilizing the more sophisticated forms of behavioural adaptation.

Of the three fundamental behavioural adaptations to cold that humans developed in prehistoric times – the use of fire, shelter, and clothing – it is the last innovation that is probably the most recent and, in terms of its repercussions, the most important from a thermal perspective. Among the reasons is that, firstly, clothing provides the most consistent physiological protection and, secondly, it is highly portable. This means that the adoption of clothing may be associated with a more prominent mismatch between morphological and environmental variation among those modern human groups where its use has been most prolonged and habitual – especially where its regular use has become increasingly mediated by cultural factors that can operate more or less independently of thermal requirements.

During the late Pleistocene, there is a marked discontinuity between Neanderthal and early modern humans in Europe in terms of limb and trunk proportions, with fully modern humans subsequently showing a temporal trend over some 20,000 years from tropical to more temperate body proportions (Holliday, 1997). The retention of tropical limb proportions among fully modern humans in Europe during the LGM might be attributable to "improved cultural buffering" – which, for reasons of thermal physiology, would include tailored, multi-layered garment assemblages (Gilligan, 2007a: 508–509) – although extraneous gene flow from Africa could also be responsible, with a subsequent shift to more cold-adapted body proportions attributable to the "harsh climate" of the LGM (Holliday, 1997: 442–444).

Femoral Neck-Shaft Angle (NSA)

The extent to which the NSA in humans varies in relation to climate (or whether, instead, its variation relates more to factors such as lifestyle and activity levels) remains an open question despite considerable research in recent years. Trinkaus (1993, 1994) interpreted the higher NSA of near-modern humans (Qafzeh-Skhul) compared to Neanderthals in the Near East as a reflection of a less mobile and physically demanding lifestyle among the former (cf. Arensburg, 1994), despite the more stocky build of Neanderthals being consistent with their having developed morphological adaptations to cooler European environments (and hence possessing a lower NSA). In supporting his conclusion, he cited NSA data on 21 modern samples from various sources grouped into three lifestyle categories (foraging, agricultural, and urban). The NSA showed a modest increase across these groups, which he interpreted as reflecting the shift to a more sedentary lifestyle rather than any differences in thermal conditions. A subsequent study employing samples from 30 modern groups reported a similar trend, and specifically discounted a climatic association after finding no significant trend between four groupings (Sub-Saharan African, European/Mediterranean, East Asian, and Native American) and climate as measured by latitude (Anderson and Trinkaus, 1998: 284). In contrast, Weaver (2003) analysed 39 anatomical markers on the hip and femur among human groups from warm and cold climates and found that Neanderthal femora (including their lower NSA) are clearly consistent with the cold-climate pattern (Figure 2). Interestingly, he suggested that the lower NSA among cold-climate groups may reflect differing developmental forces on the femoral neck secondary to climate-related differences in body proportions, rather than any purported differences due to lifestyle or activity levels (ibid.: 6928). In other words, a broader, more cold-adapted body shape will increase biomechanical loading on the femoral neck during early development, resulting in a lower NSA. He concluded that "most if not all clear contrasts in shape between the femora of Neandertals and near-modern humans seem to be secondary consequences of differences in climate-induced body proportions", although "the exact mechanical link between neck-shaft angle and body proportions needs to be established" (ibid.).

Figure 2

Differing NSA associated with warm climate (left) and cold climate (right) pelvic shape and lower limb proportions (redrawn from Weaver, 2003: 6928)

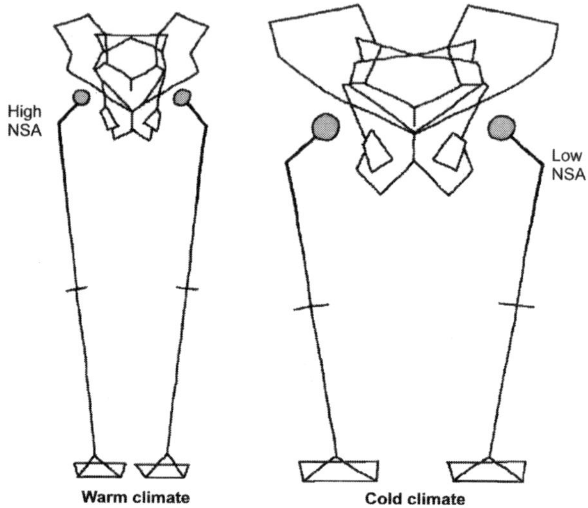

Also relevant are studies looking at osteological indicators of general femoral "robusticity" (i.e., massiveness or stockiness) in relation to climate and activity levels, with an emphasis on cross-sectional shape of the femoral shaft. Pearson (2000) found that while activity levels (hunter-gatherer vs sedentary lifestyles) probably have an influence, the evidence from a range of modern and fossil groups tends to favour a stronger relationship between robusticity and climate. Wescott and Cunningham (2006) reported an activity effect on femoral robusticity in females, though not males, associated with horticultural intensification among the Arikara (an American tribe on the Great Plains). Other studies (e.g., Weaver and Steudel-Numbers, 2005; Sládek et al., 2006; Westcott, 2006) either infer a climatic link, or fail to clearly substantiate a mobility/activity effect. Among Australian Aborigines, Collier (1989) found that midshaft diameters (antero-posterior and medio-lateral) for his Murray Valley sample (n = 120) were relatively low and "gracile", consistent with them occupying the warmest environment of the populations sampled (Inuit, American whites, Arikara Indians, and Romano-Britons from the Poundbury site in England). At the opposite climatic extreme, a study of the indigenous inhabitants of Tierra del Fuego (n = 24) revealed comparatively "robust" femoral shaft indices, consistent with a cold adapted physique, but little evidence for any effect due to differing habitual activity patterns between marine foragers (the Yahgan) and terrestrial hunter-gatherers, the Selk'nam Onas (Pearson and Millones, 2003, 2005). The complexity of possible relationships between femoral shape and both climate and activity is underscored by a study comparing the likely contributory roles of habitual activity patterns and climate on upper and lower limb robusticity among four groups – Fuegians, Native Americans, southern Africans, and Andaman Islanders (Stock, 2006). Results point to a mobility effect proximally and a climate effect distally in the upper limb, and vice versa in the lower limb (i.e., a more prominent climate effect in the proximal femur) and it is suggested that these complex effects "may be confounded by systems of cultural buffering from environmental stress" (ibid.: 200) – which presumably include clothing.

Among fossil hominins, Trinkaus and Ruff (1999) found that differences in femoral shaft geometry between Neanderthals and near-modern humans in the Near East (notably, greater medio-lateral diameters among Neanderthals) are probably related to body shape differences rather than any differences in habitual activity patterns. Conversely, Holt (2003) interpreted a shift from a more antero-posteriorly elongated to a more circular cross-section of the femoral shaft between early and late upper palaeolithic and mesolithic modern humans in Europe as reflecting a trend towards reduced mobility, with reduced cortical thickness suggesting reduced biomechanical loading. CT scanning of the archaic *Homo* tibia from Boxgrove in England, dated to the Marine Isotope Stage (MIS) 13 interglacial around 500,000 years ago, shows medio-lateral reinforcement comparable to that of Neanderthals and suggests a cold-adapted body shape despite the MIS 13 context (Trinkaus, Stringer, Ruff, Hennessy, Roberts and Parfitt, 1999a). Yet, the associated faunal data indicate a cooler, more temperate climate late in MIS 13, which may signify exposure to winter cold stress and stronger selection for biological thermal adaptations, given the likely absence of effective behavioural adaptations in the form of shelter and clothing (ibid.: 22). The late Neanderthal specimen associated with a Chṛtelperronian culture at Saint-Césaire has a femoral shaft geometry – with medio-lateral reinforcement – consistent with a cold-adapted body shape, and overall cortical thickness suggesting a high body mass even by Neanderthals standards (Trinkaus, Churchill, Ruff and Vandermeersch, 1999b), which could reflect biological response to greater cold stress in late MIS 13. Similarly, CT scans of the femora of the Tyrolean "Iceman", found at an altitude of 3,200m and dating to around 5,000 years ago, show medio-lateral strengthening suggesting a relatively stocky, cold-adapted body build (Ruff et al., 2006), which may indicate considerable cold exposure despite the presence of clothing (Spindler, 1994: 132–147).

Additionally, the femoral NSA is of considerable medical interest, as it has been implicated in the varying susceptibility of different groups to hip osteoarthritis and risk of hip fracture – a major cause of disability in the elderly and, given the aging population structure in industrialized Western societies and the expense of hip replacement surgery, a significant health cost. An early study (Walensky and O'Brien, 1968) reported that the lower NSA among American white females compared to American blacks may be a contributing factor to the higher incidence of hip fractures among whites. However, other research findings are more equivocal with respect to the NSA and suggest that reduced bone mineral density (osteoporosis) is the most significant factor – hence the heightened susceptibility to fracture among post-menopausal women (e.g., Cheng et al., 1997; Pulkkinen et al., 2006; Gnudi et al., 2007; Doherty et al., 2008).

Among modern humans the NSA varies quite widely even within small population sizes, with adult values generally in a range between 120° and 140°; more extreme values of <110° and >150° (known as coxa varus and coxa valgus, respectively) are rare but do occur. Not only is the variation marked within groups, NSA can show considerable asymmetry between left and right, although there would appear to be no consistent side difference. Whether a gender difference exists is uncertain: anatomy textbooks often state that the NSA is somewhat lower on average in females (given the broader pelvic shape), but available evidence in physical anthropology suggests that "any sexual differences are small and inconsistent" (Anderson and Trinkaus, 1998: 283). The NSA is higher in juveniles and declines during childhood, usually reaching adult values by early adolescence after which it remains stable – though there may be a minor further decline with advanced age (e.g., 80 years and over).

Measurement issues

A paucity of research on the NSA is in part attributable to the uncertainties regarding methods for determining the NSA (Figure 3). Two anatomical aspects of the proximal femur can make measurement difficult. The first relates to the highly variable and often irregular contours of the neck itself, which can render definition of the neck axis ambiguous. Even defining the shaft axis is not always straightforward as there is not uncommonly a degree of curvature in the shaft, most typically anterior bowing. The second difficulty relates to the varying amount of anterior rotation of the neck in relation to the shaft – known as anteversion of the femoral neck. This second aspect means that measuring the NSA on a standard osteometric board is not viable, and it is especially problematical where NSA measurements are taken from radiographs, since anteversion (not readily measurable on an osteometric board and not easily detectable on standard X-rays) exaggerates the NSA. While various methods and formulae have been proposed to mitigate or compensate for the consequent error (e.g., Ogata and Goldsand, 1979; Trinkaus, 1993: 398), in practice it is preferable to measure the NSA by hand and

Figure 3

The femoral neck-shaft angle (NSA) shown on the right femur (anterior view)

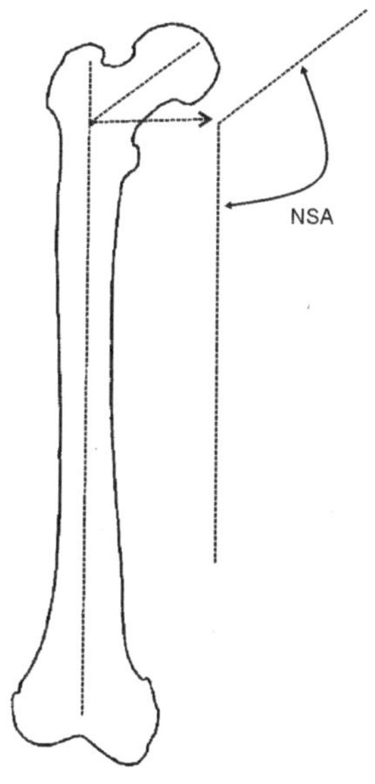

avoid using NSA data obtained from radiographs (Olsen et al., 2009). Alas, much of the older published data on NSA (still cited often in the literature) derive from radiographs, and even recent studies utilize radiographs (e.g., Igbigbi, 2003). Moreover, notwithstanding the problems of hand measurement (related essentially to anatomical variability and a lack of reliable anatomical markers), inter- and intra-observer errors have been shown to be acceptably small (usually no greater than 1–2°) in relation to differences in group means (Anderson and Trinkaus, 1998: 280). In this chapter, measurements were taken by hand on the anterior surface of the femur in a coronal plane parallel to the neck and proximal shaft, which has the advantage of removing any confounding effect of anteversion on the measurement (Trinkaus, pers.comm.).

Materials and method

Skeletal collections of modern human femora were selected on the basis of measurable NSA; complete or largely intact femora comprise the majority, although a substantial proportion of the remainder consists of only proximal segments with sufficient portions of the neck and shaft preserved to permit measurement of the NSA. Specimens with obvious deformity or pathology (such as Paget's Disease or renodelled neck fractures) were excluded; neither were juvenile nor early adolescent bones lacking epiphyseal fusion examined. Depending on the extent of preservation or completeness other measures were taken, including maximum femoral length, femoral head diameter, medio-lateral and antero-posterior shaft diameters (subtrochanteric and mid-shaft), and bicondylar breadth; side (left or right) was recorded, as well as gender and age at death (where available) – only the NSA data are discussed here. Measurements of the NSA were taken with a simple plastic 360° goniometer.

Pilot study

Prior to examining modern human skeletal collections in various institutions, a pilot study was undertaken to assess the measurement issues and also the practicalities of the research (e.g., the extent of completeness of femora and range of anatomical variation in a typical collection, and the likely time required to measure a sample of given size). This pilot study was carried out at the Congdon Anatomical Museum, Siriraj Hospital in Bangkok, Thailand. The museum houses a well-provenanced collection of very recent local material (dating from *c.* 1949), with most individual specimens identified with information on gender and age at death. The total Bangkok sample is 243 femora representing 131 individuals, comprising 112 yielding both left and right femora (n = 224), and the remaining 19 with single femora. The age range in this Bangkok sample is excellent (minimum 20 years, maximum 87 years, mean 48.5 years, standard deviation 18.2 years). It spans young adults (juveniles and adolescent specimens being excluded due to the NSA reaching its stable adult level during adolescence) to the elderly, with the latter offering the prospect of exploring whether the NSA declines slightly with advanced age. There is a male bias in the sample (70% males, 30% females), but the side (left/right) distribution is balanced (51% and 49%, respectively). The frequency distribution of NSA scores (Figure 4) approximates a normal distribution around the mean, 126°. Descriptive statistics are shown in Table 1; notable is the failure to detect significant gender or side differences in this sample. The extent of NSA variation in the Bangkok sample was marked with a range of 22°. This finding of very substantial variation in the NSA even within a single sample derived from a reasonably localized population underscores the need for adequate sampling (in terms of both geographical coverage and sample sizes).

Table 1

Descriptive statistics for NSA in the Bangkok sample

	n	*Minimum*	*Maximum*	*Mean*	*SD*
NSA	243	115°	137°	126.0°	3.65°
Gender	137				
Male	95	116°	137°	126.4°	3.88°
Female	42	122°	136°	126.3°	3.32°
Side	243				
Left	124	119°	137°	125.9°	3.36°
Right	119	115°	137°	126.1°	3.94°

Figure 4

Frequency distribution of Bangkok NSA scores

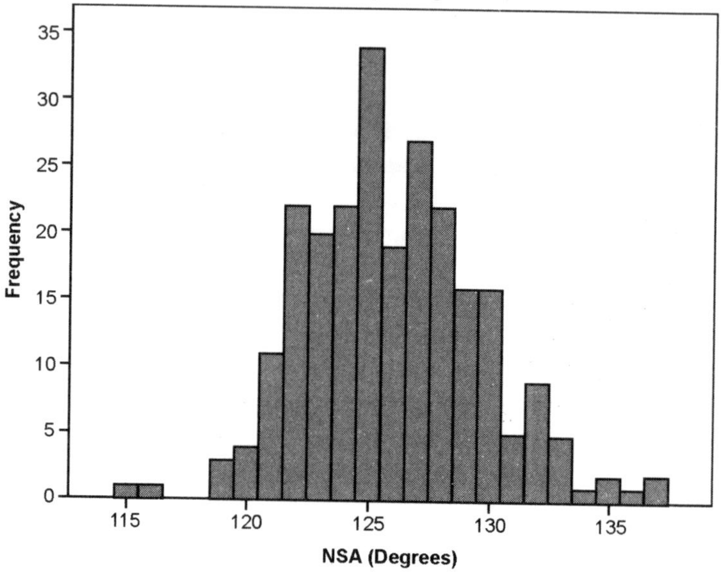

Samples

A primary purpose of this research was to obtain a useful sampling of NSA among human groups from a diverse range of environments. To this end, the quest to maximize sample size was balanced by the need for broad sampling. In some cases, very small samples were included – even a single femur from a certain locality might be measured at one institution, as this could then be added to other samples from the same or a similar locality at other institutions or combined with larger groups or regional samples in subsequent analyses. Given the requirement to be able to link each sample with geographical location and hence climatic indices, provenance to country of origin was a priority in determining inclusion – the main exceptions relate to specific human groups (such as Bushmen, Sami, and central African pygmies) where associations with modern national entities are less relevant. The numbers of femora measured and numbers of countries or groups sampled at each institution are listed in Tables 2 and 3, and the global distribution of samples is shown in Figure 5.

Figure 5

Distribution map of NSA samples (the basic 101 samples, with the USA shown as eight state sub-samples in addition to Alaska and Hawaii)

Table 2

Femoral sample sizes (n) listed alphabetically by participating institution and number of countries sampled

Institution	*Location*	*n*	*Countries*	*Other Territories*[1]	*Other Groups*[2]
American Museum of Natural History (AMNH)	New York, NY, USA	436	9	3	5
Chiang Mai University Medical School (CMU)	Chiang Mai, Thailand	214	1		
Congdon Anatomical Museum (CAM)	Bangkok, Thailand	243	1		
Florida Museum of Natural History (FLMNH)	Gainesville, FL, USA	306	1		
Institut de Paléontologie Humaine (IdPH)	Paris, France	255	5		1
Leverhulme Centre for Human Evolutionary Studies (LCHES)	Cambridge, England	350	5		4
Museé de l'Homme (MdlH)	Paris, France	2,307	60	6	5
National Museum of Nature and Science (NMNS)	Tokyo, Japan	331	1		

Cont'd...

...Cont'd

Natural History Museum (NHM)	London, England	1,991	12		6
Peabody Museum, Harvard University (PM)	Boston, MA, USA	281	8	1	1
Sapporo Medical University (SMU)	Sapporo, Japan	172	1		2
Smithsonian Institution (SI)	Washington, DC, USA	1,338	14	3	3
South Australian Museum (SAM)	Adelaide, Australia	47	1		

Notes: 1. Certain territories are listed independently of the countries responsible for their administration, e.g., Canary Islands (Spain), Easter Island (Chile), Greenland (Denmark), Guadeloupe (France), Guam (USA), Marquesas Islands (France), Nicobar Islands (India), and Tahiti (France).
2. Some groups are listed separately from their geographical or national associations, e.g., Ainu (Japan), Andaman Islanders (India), Bushmen (also known as Kung or San peoples, southern Africa), Fuegians (southern South America), Inuit (Alaska), Moriori (Chatham Island, New Zealand), Negritos (The Philippines), Okhotsk (Hokkaido, Japan), Pygmy groups (central Africa), Sami (formerly known as Lapps, northern Scandinavia and northeastern Russia), and Veddahs (Sri Lanka).

Table 3

Sample sizes (n) for each of the 101 countries/territories/groups. For statistical purposes, the one sample comprising a single femur (Cote d'Ivoire) is excluded from subsequent analyses, resulting in a total of 100 samples

Continent/ Region	*n*	*Country*	*n*	*Other Territories*	*n*	*Other Groups*	*n*
Africa	1,155	Algeria	53				
		Angola	11				
		Benin	10				
		Central African Rep.	5				
		Chad	10				
		Congo	6				
		Cote d'Ivoire	1				
		Egypt	136				
		Equatorial Guinea	2				
		Ethiopia	5				
		Gabon	25				
		Gambia	2				

Cont'd...

...Cont'd

		Ghana	2				
		Guinea	2				
		Madagascar	62				
		Mali	53				
		Morocco	59				
		Mozambique	10				
		Niger	43				
		Nigeria	2				
		Senegal	8				
		Somalia	47				
		South Africa	34				
		Sudan	130				
		Tanzania	161				
		Tunisia	2				
				Canary Islands	223		
						Bushmen	35
						Pygmy	16
Asia	1,463	Bangladesh	2				
		China	115				
		India	47				
		Indonesia	11				
		Iran	6				
		Japan	251				
		Kampuchea	2				
		Laos	2				
		Malaysia	5				
		Nepal	2				
		Pakistan	4				
		Sri Lanka	16				
		Syria	15				
		Thailand	457				
		The Philippines	92				
		Turkmenistan	4				
		Vietnam	16				
				Nicobar Islands	12		
				Russia (Siberia)[3]	18		
						Ainu	233

Cont'd...

...Cont'd

						Andamans	81
						Negrito	37
						Okhotsk	19
						Veddah	16
Australia	47	(South) Australia	47				
Europe	2,799	Armenia	2				
		Belgium	6				
		Cyprus	2				
		Czech Republic	56				
		England	1,271				
		Estonia	4				
		Finland	2				
		France	768				
		Iceland	146				
		Italy	10				
		Norway	4				
		Poland	8				
		Russia	46				
		Scotland	434				
		Spain	14				
		Sweden	2				
		Ukraine	10				
		Wales	2				
						Sami	12
North America	1,766	Canada	272				
		Dominican Republic	24				
		Mexico	65				
		USA	987				
				Greenland	90		
				Guadeloupe	4		
						Inuit	324
Pacific	332	Fiji	3				
		Kiribati	4				
		New Zealand	20				
		Papua New Guinea	59				
		Solomon Islands	20				
		Tonga	4				

Cont'd...

...Cont'd

		Vanuatu	76				
				Easter Island	94		
				Guam	2		
				Hawaii[4]	12		
				Marquesas	2		
				Tahiti	11		
						Moriori	25
South America	709	Argentina	108				
		Bolivia	33				
		Brazil	9				
		Chile	49				
		Ecuador	193				
		Peru	211				
		Surinam	4				
		Venezuela	68				
						Fuegian	34

Notes: 3. The Russian Far East (a Russian federal district formerly known loosely as Siberia) is here considered separately as part of the continent of Asia rather than Europe.

4. Hawaii, a state of the USA, is listed in the Pacific region.

A total of 8,271 femora were measured. When grouped according to country (or territory) of origin, the total number of samples is 101 (80 countries, 10 territories, and 11 other groups). Multiple regional sub-samples were also obtained for some countries (e.g., 10 state samples from the USA, and 21 provincial samples from France). One of the samples (from the Cote d'Ivoire) comprises only a single femur, which is included in the continental analyses as part of the African sample, but is excluded from the global analyses which utilize group means (yielding a net sample size of 100 countries/territories/other groups for the initial global analysis). The mean, minimum, and maximum NSA values (and standard deviations) for the 101 samples are listed in Table 4.

Table 4

Minimum, maximum, and mean NSA values (and the standard deviation) for each sample

Groups	*n*	*Minimum*	*Maximum*	*Mean*	*SD*
Africa					
Algeria	53	115	131	123.4	4.1
Angola	11	122	133	128.7	3.2
Benin	10	122	138	130.1	4.6
Canary Islands	223	110	140	125.7	5.7
Central African Rep.	5	122	135	128.2	5.3
Chad	10	124	138	132.3	4.2
Congo	6	113	138	123.7	10.4
Cote d'Ivoire	1	128	128	128.0	0.0
Egypt	136	112	140	127.0	5.0
Equatorial Guinea	2	123	131	127.0	5.7
Ethiopia	5	124	136	132.2	4.8
Gabon	25	119	138	130.0	5.1
Gambia	2	143	145	144.0	1.4
Ghana	2	125	126	125.5	0.7
Guinea	2	132	135	133.5	2.1
Madagascar	62	114	136	126.1	4.9
Mali	53	122	142	130.8	5.1
Morocco	59	114	130	120.0	3.4
Mozambique	10	112	126	120.8	3.7
Niger	43	117	140	129.0	5.1
Nigeria	2	129	130	129.5	0.7
Senegal	8	127	143	132.5	5.2
Somalia	47	117	139	125.5	4.7
South Africa	34	114	142	129.7	7.5
Sudan	130	118	143	130.5	4.4
Tanzania	161	117	142	127.8	5.0
Tunisia	2	125	126	125.5	0.7
Asia					
Bangladesh	2	126	126	126.0	0.0
China	115	113	139	127.2	5.0

Cont'd...

...Cont'd

India	47	115	140	129.9	5.0
Indonesia	11	125	136	131.6	3.2
Iran	6	114	131	125.2	6.2
Japan	251	113	140	127.2	4.6
Kampuchea	2	126	129	127.5	2.1
Laos	2	128	130	129.0	1.4
Malaysia	5	127	139	132.8	5.0
Nepal	2	126	130	128.0	2.8
Nicobar Islands	12	122	142	131.6	5.5
Pakistan	4	122	129	126.3	3.1
Siberia	18	120	136	130.0	4.1
Sri Lanka	16	127	136	129.7	2.7
Syria	15	115	134	125.3	6.2
Thailand	457	114	141	127.5	4.8
The Philippines	92	113	141	128.6	4.5
Turkestan	4	119	124	121.5	2.1
Vietnam	16	116	141	126.3	7.8
Australia					
(South) Australia	47	119	139	130.2	4.8
Europe					
Armenia	2	128	129	128.5	0.7
Belgium	6	117	132	125.3	6.6
Cyprus	2	116	121	118.5	3.5
Czech Republic	56	116	139	126.8	5.0
England	1,271	108	142	125.5	5.5
Estonia	4	123	130	126.0	3.3
Finland	2	121	122	121.5	0.7
France	768	112	142	127.0	5.3
Iceland	146	114	138	126.6	4.2
Italy	10	119	134	127.7	4.3
Norway	4	122	132	126.0	4.2
Poland	8	116	131	125.3	5.6
Russia	46	111	135	126.0	5.4
Scotland	434	116	142	127.7	4.9
Spain	14	112	131	124.2	5.0

Cont'd...

...Cont'd

Sweden	2	121	123	122.0	1.4
Ukraine	10	117	132	124.8	4.9
Wales	2	125	126	125.5	0.7
North America					
Canada	272	113	135	124.2	4.3
Dominican Republic	24	118	136	124.3	4.0
Greenland	90	118	141	129.3	4.2
Guadeloupe	4	115	123	120.8	3.9
Mexico	65	112	138	126.5	5.2
USA	987	109	142	125.6	5.3
Pacific					
Easter Island	94	117	135	126.7	3.9
Fiji	3	130	136	133.3	3.1
Guam	2	126	131	128.5	3.5
Hawaii	12	126	142	132.9	4.5
Kiribati	4	125	132	128.3	3.3
Marquesas	2	125	127	126.0	1.4
New Zealand	20	116	131	124.3	4.9
Papua New Guinea	59	121	142	132.0	4.4
Solomon Islands	20	124	136	132.4	2.9
Tahiti	11	125	140	132.5	4.6
Tonga	4	115	143	129.5	14.5
Vanuatu	76	123	141	130.9	3.8
South America					
Argentina	108	114	143	128.9	4.8
Bolivia	33	105	126	117.6	4.3
Brazil	9	126	130	128.1	1.4
Chile	49	105	136	124.5	5.6
Ecuador	193	112	144	125.9	4.9
Peru	211	109	136	122.2	5.4
Surinam	4	122	131	127.3	3.9
Venezuela	68	115	141	127.0	6.0
Other Groups					
Ainu	233	112	136	125.3	3.9

Cont'd...

...Cont'd

Andaman Islands	81	125	148	136.1	4.4
Bushmen	35	120	131	126.7	2.7
Fuegian	34	112	128	122.4	3.8
Inuit	324	109	131	120.2	4.1
Moriori	25	120	141	130.9	4.7
Negrito	37	125	137	130.9	3.6
Okhotsk	19	118	132	124.8	4.0
Pygmy	16	125	139	131.6	3.9
Sami	12	121	128	125.3	2.2
Veddah	16	126	139	132.3	3.7

Climatic indices

Each sample was assigned a local weather station with adequate meteorological records to provide (in addition to latitude) three climatic indices:

1. annual mean temperature,
2. summer mean monthly temperature (January or July, depending on hemisphere), and
3. winter mean monthly temperature (January or July).

Table 5

Descriptive statistics for the global NSA based on the whole NSA database

Sample	*n*	*Mean*	*Minimum*	*Maximum*	*SD*
Total femora	8,271	126.4	105	148	5.57
Group samples	100	127.4	117.6	144.0	3.92

A number of internet websites now offer well-sourced data for thousands of locations worldwide; the Weatherbase^{SM} site (www.weatherbase.com) was chosen for these analyses. The selected weather station, latitude, and temperature data for each sample are listed in Table 6.

Table 6

List of each sample group (and mean NSA) with the associated climate data from Weatherbase^{SM} (www.weatherbase.com). The selected weather station is typically the designated capital city (or, in the case of the Andaman Islands for instance, the recognized administrative centre) of the country/state/territory. Latitude is that of the weather station, with climate data comprising mean annual temperature (Annual), and mean January or July monthly temperatures (°C) for Summer and Winter (depending on hemisphere)

Groups	*NSA*	*Weather Station*	*Latitude*	*Annual*	*Summer*	*Winter*
Africa						
Algeria	123.4	Algiers	37	17	25	11
Angola	128.7	Luanda	9	24	25	20
Benin	130.1	Porto-Novo	6	26	25	27
Canary Islands	125.7	Las Palmas	28	20	23	17
Central African Rep.	128.2	Bangui	4	26	25	24
Chad	132.3	N'jamena	12	28	28	23
Congo	123.7	Kinshasa	4	26	26	23
Cote d'Ivoire	128.0	Yamoussoukro	7	26	25	26
Egypt	127.0	Cairo	30	21	28	13
Equatorial Guinea	127.0	Malabo	4	25	24	24
Ethiopia	132.2	Addis Ababa	9	16	15	16
Gabon	130.0	Libreville	0	26	27	24
Gambia	144.0	Banjul	13	26	27	23
Ghana	125.5	Accra	6	27	25	27
Guinea	133.5	Conakry	10	27	25	26
Madagascar	126.1	Antanarivo	19	18	21	15
Mali	130.8	Bamako	13	28	26	25
Morocco	120.0	Rabat	34	17	22	12
Mozambique	120.8	Maputo	26	23	27	19
Niger	129.0	Niamey	13	29	28	24
Nigeria	129.5	Abuja	9	23	25	28
Senegal	132.5	Dakar	15	24	27	21
Somalia	125.5	Mogadishu	2	27	26	27
South Africa	129.7	Pretoria	26	18	23	12

Cont'd...

...Cont'd

Sudan	130.5	Khartoum	16	30	32	23
Tanzania	127.8	Dodoma	6	22	23	19
Tunisia	125.5	Tunis	37	18	26	11
Asia						
Bangladesh	126.0	Dhaka	24	26	28	19
China	127.2	Beijing	40	12	26	–3
India	129.9	Delhi	29	25	30	14
Indonesia	131.6	Jakarta	6	27	26	27
Iran	125.2	Tehran	36	17	31	1
Kampuchea	127.5	Phnom Penh	12	27	28	26
Japan	127.2	Tokyo	36	15	25	5
Laos	129.0	Vientiane	18	26	28	22
Malaysia	132.8	Kuala Lumpur	3	27	27	27
Nepal	128.0	Kathmandu	28	18	23	10
Nicobar Islands	131.6	Port Blair	12	27	27	26
Pakistan	126.3	Islamabad	34	21	29	10
Siberia	130.0	Vladivostok	43	5	17	–12
Sri Lanka	129.7	Colombo	7	27	28	27
Syria	125.3	Damascus	33	16	26	6
Thailand	127.5	Bangkok	14	28	29	26
The Philippines	128.6	Manila	15	27	27	26
Turkmenistan	121.5	Ashgabat	38	16	30	5
Vietnam	126.3	Hanoi	21	24	30	16
Australia						
(South) Australia	130.2	Adelaide	35	16	21	13
Europe						
Armenia	128.5	Yerevan	40	12	26	–2
Belgium	125.3	Brussels	51	10	17	3
Cyprus	118.5	Nicosia	35	18	28	9
Czech Republic	126.8	Prague	50	8	17	–1
England	125.5	London	51	10	16	3
Estonia	126.0	Tallinn	59	5	16	–3
Finland	121.5	Helsinki	60	5	16	–5
France	127.0	Paris	49	11	19	3

Cont'd...

...Cont'd

Iceland	126.6	Reykjavik	64	4	11	–1
Italy	127.7	Rome	42	15	23	8
Norway	126.0	Oslo	60	6	17	–3
Poland	125.3	Warsaw	52	8	17	–1
Russia	126.0	Moscow	56	4	17	–8
Scotland	127.7	Edinburgh	56	8	15	3
Spain	124.2	Madrid	40	13	24	5
Sweden	122.0	Stockholm	60	6	17	–2
Ukraine	124.8	Kiev	50	7	18	–6
Wales	125.5	Cardiff	51	10	16	5
North America						
Canada	124.2	Ottawa	45	6	21	–10
Dominican Republic	124.3	Santo Domingo	18	26	27	24
Greenland	129.3	Nuuk	64	–1	7	–8
Guadeloupe	120.8	Basse-Terre	16	26	27	24
Mexico	126.5	Mexico City	19	17	18	13
USA	125.6	Washington	39	14	26	2
Pacific						
Easter Island	126.7	Hanga Roa	27	21	23	18
Fiji	133.3	Suva	18	24	26	23
Guam	128.5	Hagatna	13	27	27	26
Hawaii	132.9	Honolulu	21	25	26	22
Kiribati	128.3	Tarawa	1	28	28	28
Marquesas	126.0	Atuona	10	26	27	25
New Zealand	124.3	Wellington	41	12	17	8
Papua New Guinea	132.0	Port Moresby	9	27	27	25
Solomon Islands	132.4	Honiara	9	26	27	26
Tahiti	132.5	Papeete	18	26	26	25
Tonga	129.5	Nuku'alofa	21	23	25	21
Vanuatu	130.9	Port Vila	18	24	26	22
South America						
Argentina	128.9	Buenos Aires	35	16	23	10
Bolivia	117.6	Sucre	19	15	16	13
Brazil	128.1	Brasilia	16	21	22	18

Cont'd...

...Cont'd

Chile	124.5	Santiago	33	14	21	8
Ecuador	125.9	Quito	0	14	14	14
Peru	122.2	Lima	12	20	23	17
Surinam	127.3	Paramaribo	5	27	27	26
Venezuela	127.0	Caracas	11	23	23	21
Other Groups						
Ainu	125.3	Sapporo	43	8	21	–5
Andaman Islands	136.1	Port Blair	12	27	27	26
Bushmen	126.7	Gaborone	25	21	26	13
Fuegian	122.4	Punta Arenas	53	6	10	1
Inuit	120.2	Juneau	58	5	13	–4
Moriori	130.9	Waitangi	44	10	14	7
Negrito	130.9	Manila	15	27	27	26
Okhotsk	124.8	Sapporo	43	8	21	–5
Pygmy	131.6	Kinshasa	4	26	26	23
Sami	125.3	Helsinki	60	5	16	–5
Veddah	132.3	Colombo	7	27	28	27

Results

Mean global NSA

There are two methods for calculating the global mean NSA using this database: the first uses total number of individual femora (n = 8,271), and the second uses mean NSA scores for the group samples (n = 100, with the one group sample comprising a single femur – the Cote d'Ivoire – being excluded). The mean NSA scores are 126.4° and 127.4°, respectively; descriptive statistics are shown in Table 5, and the NSA scores for individual femora are plotted (against a normal distribution curve) in Figure 6.

Global climatic trends in the NSA

As an initial statistical procedure for ascertaining any overall climatic trends across the full database (Gilligan, 2010a), Pearson correlation coefficients were calculated on the mean NSA for each of the100

Figure 6

Frequency distribution of the NSA in the global sample (n = 8,271)

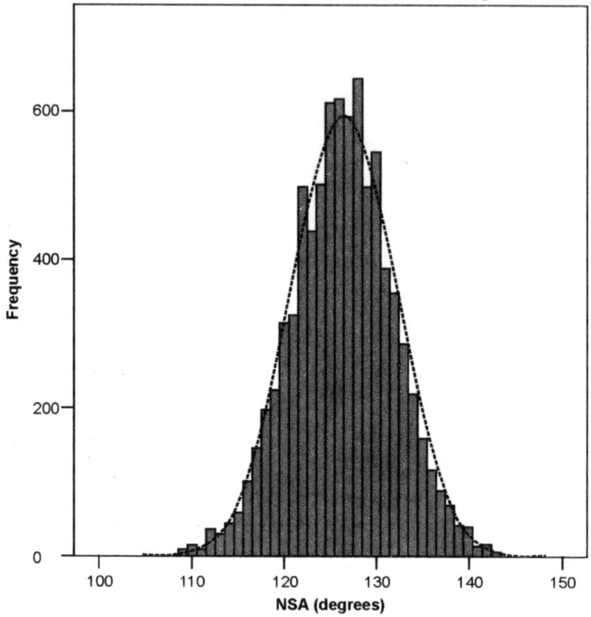

samples – the one case (Cote d'Ivoire) comprising a single femur being excluded. The four environmental variables are latitude, mean annual temperature, and the two seasonal means (summer and winter). Tests were carried out with SPSS® Graduate Pack 16.0 for Windows.

Table 7

Results for the 100 NSA samples: Pearson correlation coefficients (r) for latitude and the climate variables, significance level of the result (two–tailed tests), and an estimate (r^2) of total variance (%) in NSA attributable to the variables

Variable	*n*	*r*	*Significance (**.005, 2-tailed)*	r^2
Latitude (°N/S)	100	–.430	.000**	18.5%
Annual mean temperature (°C)	100	+.457	.000**	20.9%
Summer mean monthly temperature (°C)	100	+.300	.002**	9%
Winter mean monthly temperature (°C)	100	+.461	.000**	21.3%

Table 7 shows the Pearson correlation coefficients (r values) between NSA and the four environmental indices, together with significance level (two-tailed tests) and the r^2 value (which estimates the proportion of variance in NSA that may be attributed to each environmental variable). A scatterplot of group NSA scores and mean winter temperature is shown in Figure 7.

Figure 7

Scatterplot of group mean NSA scores and winter monthly temperature (January/July)

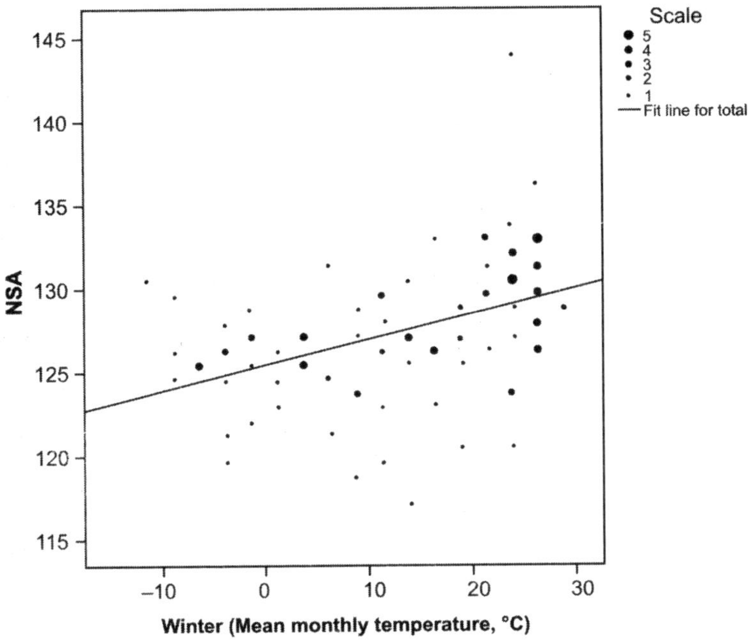

With the exception of the summer mean monthly variable (p = .002), all correlations are significant at the .000 level, with winter mean monthly temperature yielding the highest r value. While r^2 values are modest (~20% for latitude, annual and winter temperatures), these results suggest that the observed associations between NSA and environmental variables are anthropologically (as opposed to merely statistically) significant, given the variability in NSA even within small samples.

Continental means and trends

The global climatic patterning revealed in the preceding analysis can now be explored in more detail. With the exception of Australia (which is represented only by the single small sample from Roonka), the database includes reasonably extensive coverage at the continental level. In addition, the large samples from two countries (the USA and France) include multiple sub-groups provenanced to local regions, allowing for the exploration of possible climatic relationships at a more refined level.

Continental means

The mean NSA figures from each of the continents – counting the Pacific as a "continent" – are presented in Table 8 and Figure 8. Each of these continental means is calculated in two ways: first, using the individual NSA of each femur (that is, independent of the countries, territories, and other groups) and, second, using the group means. The only continent where the resulting mean NSA values differ to any extent is Europe, where approximately half of the continental NSA derives from a single large sample (the Poundbury sample from England), and that difference is small (1°).

Figure 8

Continental distributions of mean NSA (100 group means)

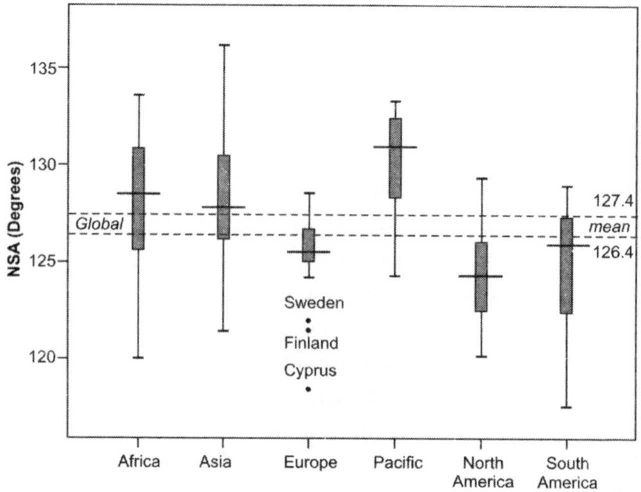

Table 8

Continental mean NSA values, calculated on both the total femoral samples and on the group NSA means

Continent	*n*	*Mean*	*Minimum*	*Maximum*	*SD*
Africa					
Total	1,155	127.2	110	145	5.72
Samples	28	128.5	120	144	4.60
Asia					
Total	1,463	128.8	112	148	5.31
Samples	23	128.5	122	136	3.29
Europe					
Total	2,799	126.3	108	142	5.32
Samples	19	125.3	119	129	2.39
North America					
Total	1,766	124.6	109	142	5.38
Samples	7	124.4	120	129	3.18
Pacific					
Total	332	129.9	115	143	4.99
Samples	13	129.6	124	133	2.91
South America					
Total	709	124.7	105	144	5.81
Samples	9	124.9	118	129	3.61

The highest NSA mean is found in the Pacific, with high values also in Africa and, to a lesser degree, Asia; lower continental means are found in the Americas, with Europe occupying an intermediate position. Even at this gross level, it is possible to suggest that these continental differences could reflect the differing thermal environments (and the environmental histories) of the populations. The Pacific samples comprise approximately half Melanesian and half Polynesian groups (Micronesia being represented only by two very small samples, from Kiribati and Guam). Melanesians have a long history (and prehistory extending at least into the late Pleistocene) in the tropics, while the Polynesians have their mid-Holocene origins in Melanesia and south-eastern Asia. The reasonably high African mean is expected given that tropical and subtropical parts of the continent are well represented. The similar Asian mean may result from a majority (14/24) of the samples deriving from the tropics, with only a quarter (6/24) deriving from cooler temperate environmental zones – China (a single Smithsonian sample provenanced loosely to northern China), Japan, Nepal, the Russian Far East and the Ainu and Okhotsk groups. An extensive use of thermally, effective clothing among most if not all of these latter six groups might ameliorate any climatic influence: all have relatively high NSA values (the Russian Far East – formerly known as Siberia – having the highest, 130°), with the exception of the Okhotsk (mean NSA 124.8°) and, to a lesser extent, the Ainu (125.3°) groups. Relatively, lower NSA group means in Europe are expected on thermal grounds, as is a narrower NSA range compared to other continents, reflecting the more restricted geographical range in terms of latitude (and hence a more restricted range of temperatures). While the Cyprus outlier is anomalous in this context, it may represent an unreliable mean population estimate given the very small sample size ($n = 2$), whereas the two Scandinavian outliers are consistent with thermal expectations.

The lower NSA means for the Americas are of interest in terms of a likely climatic influence, given the late Pleistocene origins of the Native American populations in north-eastern Asia. Even with archaeological evidence for the use of complex clothing among the early immigrants to the New World (Gilligan, 2010: 44), a degree of climatic selection for morphological cold adaptations among the ancestors of Native Americans in polar regions at the end of the last ice age is not unexpected.

Continental climate trends

The group mean NSA values on each continent are analysed here in relation to the climatic indices, to examine for the existence of climatic trends in NSA at the continental level.

Africa

Pearson correlations between group NSA and the climate variables (and the results of significance tests) for the 28 African groups are shown in Table 9. The correlations are modest but nonetheless follow anticipated climatic trends (negative for latitude, positive for temperature), with only mean summer temperature failing to reach significance at the 0.05 level. The results for mean winter temperature are plotted in Figure 9.

Table 9

NSA and climate correlations for Africa

	Climate Variable/Proxy	*N*	*r*	*Significance (*.05, 1-tailed)*
L	Latitude (°N/S)	28	−.381	.045*
A	Annual mean monthly temperature (°C)	28	+.390	.020*
S	Summer (July) mean temperature (°C)	28	+.140	.239
W	Winter (January) mean temperature (°C)	28	+.377	.024*

Figure 9

Scatterplot of NSA and winter temperature for Africa

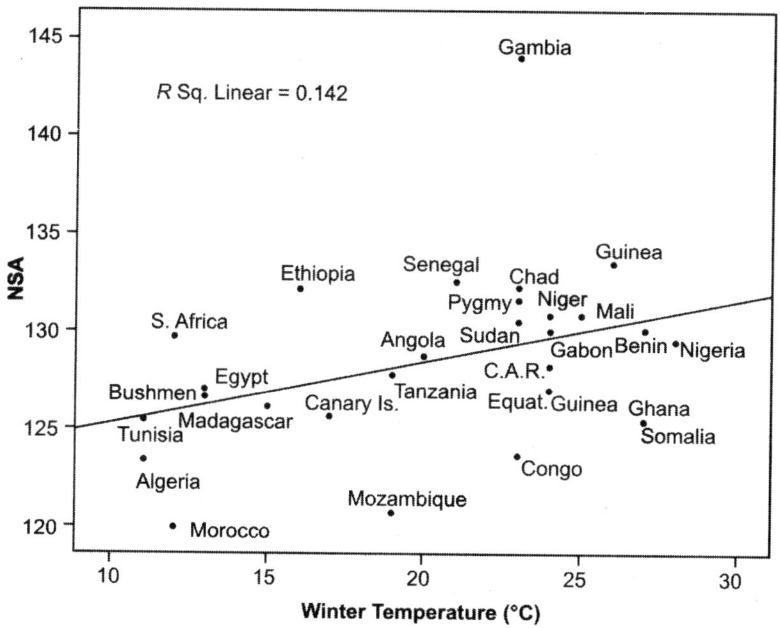

The very small ($n = 2$) sample from Gambia emerges as an outlier with its high NSA of 144° but, even so, the distribution of these African group means follows an overall trend with winter temperature. Morphological adaptation to seasonal temperature variation may be limited due to the overriding influence of an ancient hominin adaptation to heat stress in Africa which, even in temperate parts of the continent, continues to the present in terms of generally high summer temperatures. Relatively little variation in mean summer temperatures across the continent (Table 6) compared to the variation in mean annual and winter temperatures may account for a poor correlation between NSA and mean summer temperature.

Asia

Greater temperature ranges and seasonality in the climatic indices (and also a greater latitudinal range) contribute to stronger correlations for the NSA in Asia compared to Africa (Table 10). Summer temperature again provides the weakest correlation (and is slightly negative), whereas a good correlation with winter temperature is found. The group distributions are plotted against latitude in Figure 10, and show a consistent trend with this climate proxy. Two groups show unusually high NSA for their latitude (although neither could be termed outliers): the Russian Far East (mentioned above) and the Andaman Islands. The latter is a group of special interest here, given the absence of clothing prior to British settlement (Gilligan, 2010: 28). The high NSA (136.1°, n = 81) presumably reflects morphological adaptation to a quite uniformly warm thermal environment (Table 6): mean annual and summer temperatures of 27°C, and mean winter temperature 26°C.

Table 10

NSA and climate correlations for Asia

	Climate Variable/Proxy	*N*	*r*	*Significance (**.005, 1-tailed)*
L	Latitude (°N/S)	23	–.738	.000**
A	Annual mean monthly temperature (°C)	23	+.574	.002**
S	Summer (July) mean temperature (°C)	23	–.007	.488
W	Winter (January) mean temperature (°C)	23	+.644	.000**

Figure 10

Scatterplot of NSA and latitude for Asia

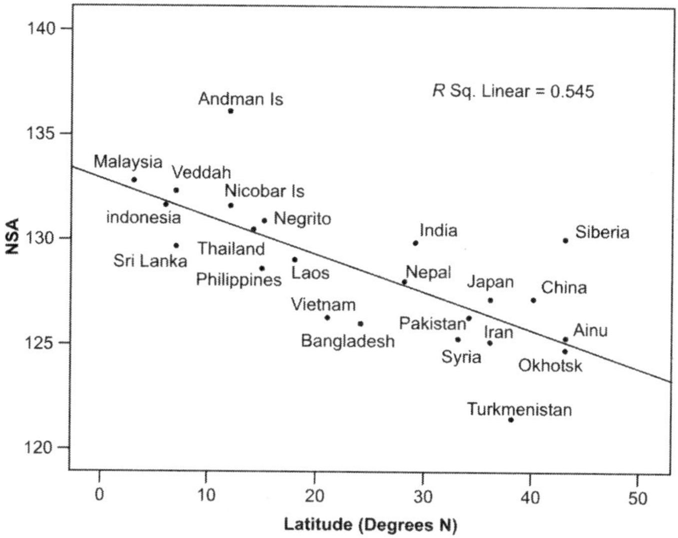

Europe

The NSA correlations for Europe (Table 11) are very low, and they occur in the opposite direction to that anticipated – positive for latitude, negative for temperature. The group distributions are plotted against latitude in Figure 11, which shows Cyprus as an extreme outlier (NSA 118.5°). Given the very small sample size for Cyprus (n = 2) and the anomalous NSA value, the correlations and statistical tests for Europe were repeated with Cyprus omitted (Table 12).

Table 11

NSA and climate correlations for Europe

Climate Variable/Proxy		*N*	*r*	*Significance (*.05, 1-tailed)*
L	Latitude (°N/S)	19	+.116	.318
A	Annual mean monthly temperature (°C)	19	–.184	.226
S	Summer (July) mean temperature (°C)	19	–.256	.145
W	Winter (January) mean temperature (°C)	19	–.116	.317

Figure 11

Scatterplot of NSA and latitude for Europe

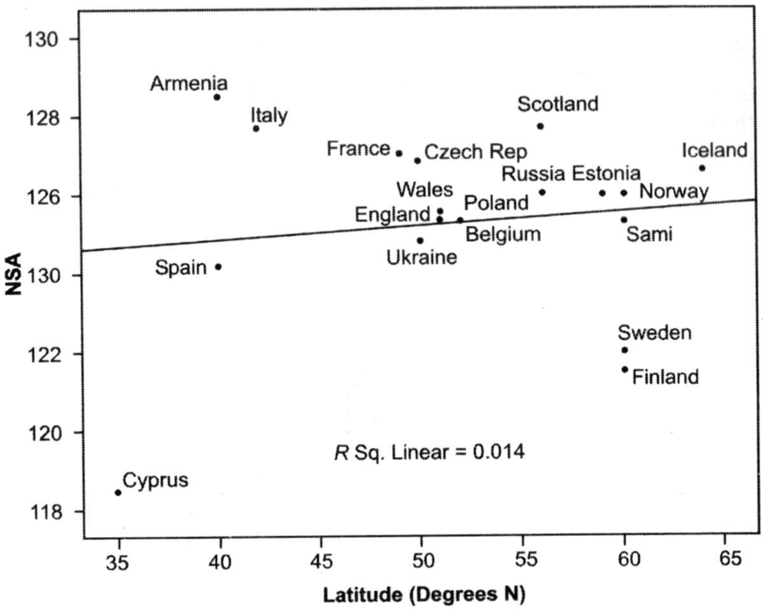

Table 12

NSA and climate correlations in Europe without the Cyprus sample

	Climate Variable/Proxy	*N*	*r*	*Significance (*.05, 1-tailed)*
L	Latitude (°N/S)	18	–.366	.067
A	Annual mean monthly temperature (°C)	18	+.364	.069
S	Summer (July) mean temperature (°C)	18	+.228	.182
W	Winter (January) mean temperature (°C)	18	+.297	.116

Without the Cyprus sample, the NSA correlations revert to the anticipated pattern (negative for latitude, positive for temperature), although they remain modest and none reach statistical significance. Assuming that sampling is adequate and that the group means are representative of the respective populations, the question arises as to why climate trends for the NSA in Europe might be relatively minor by other continental standards (Figure 12). One reason may be the comparatively small latitudinal (and hence climatic) range: approximately 25°, compared to ~50° for Asia and ~60° for Africa. Another possible factor affecting any thermal influence on the NSA in Europe may relate to the generally ubiquitous use of thermally effective (i.e., complex) clothing throughout the continent, which would tend to ameliorate the influence of the external thermal environment.

Figure 12

Scatterplot of NSA and latitude in Europe without the Cyprus sample

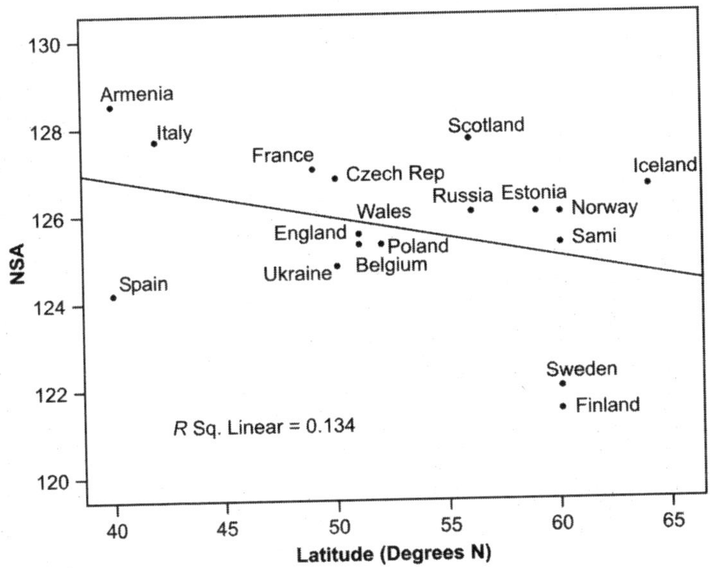

North America

The correlation results for North America are shown in Table 13 and the group distributions for latitude are plotted in Figure 13. The correlations are very low and, for mean annual and summer temperatures, occur in the opposite direction to that anticipated on thermal grounds. The number of sample groups (five) for North America is small and, in terms of trends expected on thermal grounds, it is evident that Greenland (NSA 129.3°, $n = 90$) is an anomalous outlier.

Figure 13

Scatterplot of NSA and latitude in North America

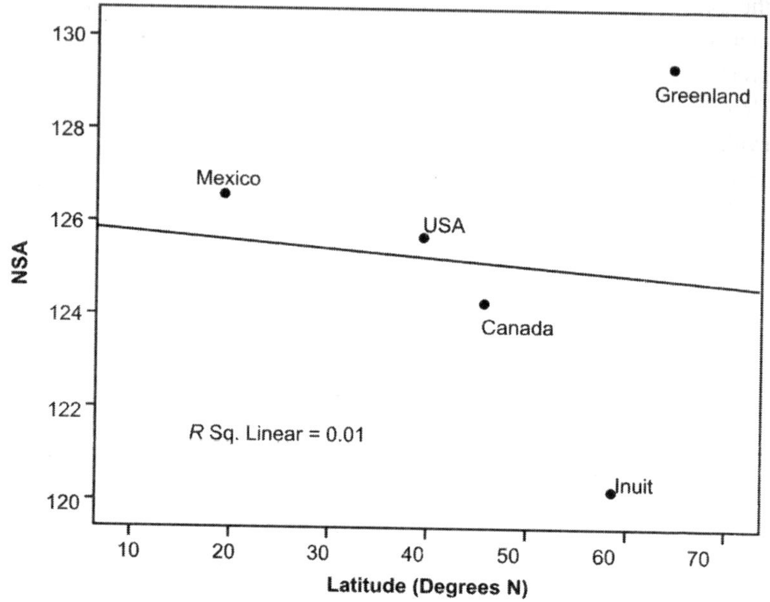

Table 13

NSA and climate correlations for North America

Climate Variable/Proxy		*N*	*r*	*Significance (*.05, 1-tailed)*
L	Latitude (°N/S)	5	–.099	.437
A	Annual mean monthly temperature (°C)	5	–.059	.462
S	Summer (July) mean temperature (°C)	5	–.206	.370
W	Winter (January) mean temperature (°C)	5	+.118	.425

Possible reasons for the unexpectedly high NSA for Greenland are not readily apparent: the sample is of adequate size and consists of reasonably well-provenanced collections in three institutions (the SI, MdlH, and PM) – all designated as "Eskimos", provenanced mainly to Nuussuaq in western Greenland, and Angmagssalik and Nanortalik in the south, respectively. Whether effective "cultural buffering" from the climate could play a role is unclear: the comparable Inuit sample follows the expected climatic trend (although the latter groups, derived from a maritime context on the Aleutian Islands, were probably exposed to greater wind chill stresses). When the analyses are repeated with Greenland omitted (Table 14), the expected climatic trends become evident, although the small number of samples (four) mitigate against the strong correlations reaching statistical significance. The North American group results for latitude (without Greenland) are plotted in Figure 14.

Figure 14

Scatterplot of NSA and latitude for North America without the Greenland sample

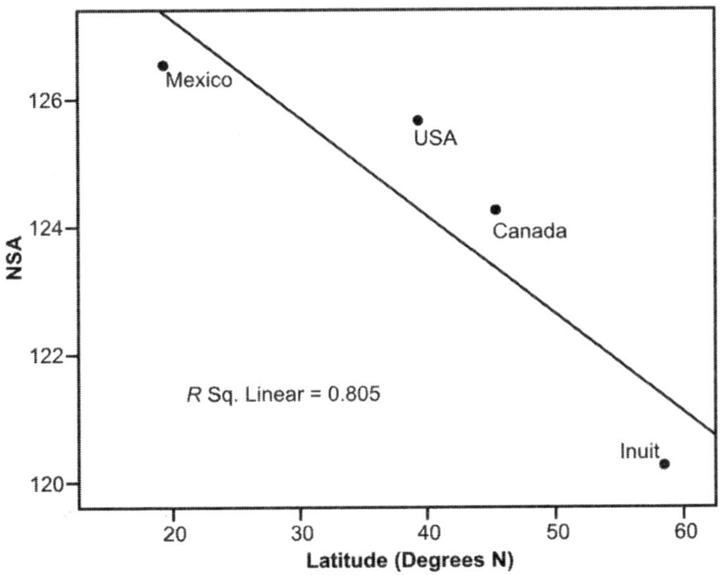

Table 14

North American NSA/climate correlations without Greenland

	Climate Variable/Proxy	*N*	*r*	*Significance (*.05, 1-tailed)*
L	Latitude (°N/S)	4	–.897	.051
A	Annual mean monthly temperature (°C)	4	+.847	.076
S	Summer (July) mean temperature (°C)	4	+.696	.152
W	Winter (January) mean temperature (°C)	4	+.596	.202

Pacific

Correlation results for the Pacific region are shown in Table 15, and the distribution of group means against latitude is shown in Figure 15. The correlations are modest, and it is evident that one sample – the Moriori (NSA 130.9°, $n = 25$) – appears to occupy an anomalous position with a high NSA given the latitude (44°S, corresponding to the weather station at Waitangi in the Chatham Islands, the ancestral home of the Moriori off the eastern coast of New Zealand).

Figure 15

Scatterplot of NSA and latitude for the Pacific groups

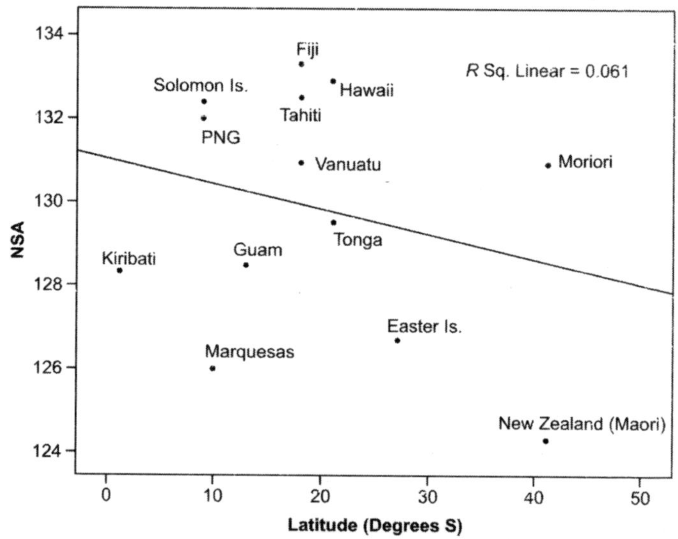

Table 15

NSA and climate correlation results for the Pacific

Climate Variable/Proxy		*N*	*r*	*Significance (*.05, 1–tailed)*
L	Latitude (°N/S)	13	–.248	.207
A	Annual mean monthly temperature (°C)	13	+.357	.116
S	Summer (July) mean temperature (°C)	13	+.368	.108
W	Winter (January) mean temperature (°C)	13	+.346	.123

As with the Greenland sample, reasons for the Moriori being an outlier are not readily apparent: while the total sample size is not large ($n = 25$), it derives from four institutions (the AMNH, PM, NHM, and LCHES), with the Moriori NSA being high in each of these four collections. The comparable New Zealand sample, the Maori (NSA 124.3°), follows an anticipated thermal trend and, given the cooler and more exposed environment of the Chatham Islands (with slightly lower mean temperatures, as shown in Table 6), the Moriori might be expected to have a lower, not higher, mean NSA on thermal grounds. There is not any likely difference between these two groups in terms of "cultural buffering" from the climate – the Moriori, in fact, may have used less thermally effective clothing (with no woven garments, in contrast to the Maori) and some historical accounts suggest the males sometimes dispensed with clothing altogether. Unless the ancestors of the Moriori arrived in the region from a warmer climate more recently than the Maori (for which there is no evidence), their high NSA is unexpected.

Correlations between NSA and climate for the Pacific omitting the Moriori are shown in Table 16, and the group distributions for summer temperature are plotted in Figure 16. Correlations occur in the expected directions (negative for latitude, positive for temperature) and reach statistical significance ($p \leq 0.05$) on all three temperature variables. The group means for NSA are distributed in a reasonably logical fashion given the geographical locations, although the Marquesas and the two small Micronesian samples (Kiribati and Guam) have slightly lower NSA means than might be expected.

Table 16

Pacific NSA/climate correlations without Moriori

Climate Variable/Proxy		*N*	*r*	*Significance (*.05, 1-tailed)*
L	Latitude (°N/S)	12	–.373	.116
A	Annual mean monthly temperature (°C)	12	+.553	.031*
S	Summer (July) mean temperature (°C)	12	+.570	.026*
W	Winter (January) mean temperature (°C)	12	+.529	.038*

Figure 16

Scatterplot of NSA and summer temperature for the Pacific without the Moriori

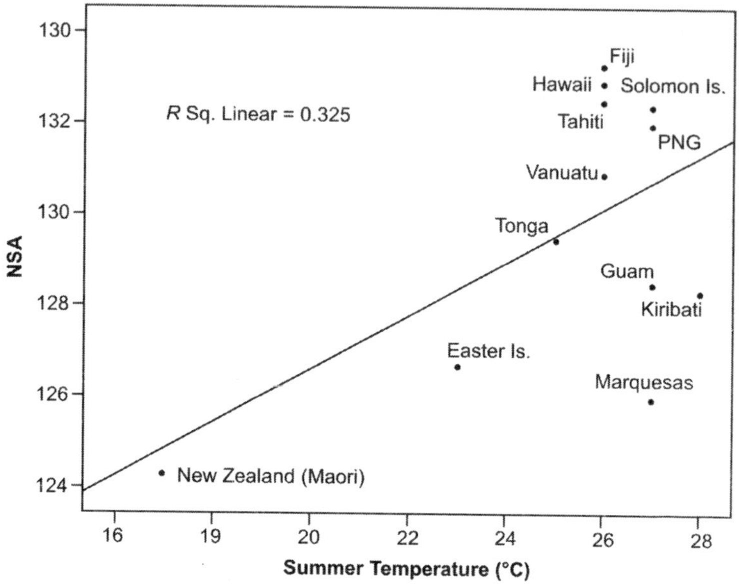

South America

Correlation results for South America are given in Table 17 and the group distributions of mean NSA against annual mean temperature are shown in Figure 17. Correlations occur in the expected directions, and while the correlation with latitude is weak, correlations on the temperature variables are modest. None reach statistical significance, although this may be partially attributable to the small number of samples in the analyses (nine groups). The distribution of the groups follows expected thermal trends, however, with the Fuegian sample (mean NSA 122.4°, $n = 34$) falling towards one extreme. A notable outlier is the Bolivian sample (mean NSA 117.6°, $n = 33$) which has the lowest South American NSA here. The sample derives from two collections (at the MdlH and PM), both having low NSA means and with a large proportion ($n = 25$) provenanced to the Lake Titicaca area – at an altitude of ~3800 m.

Figure 17

Scatterplot of NSA group means for South America

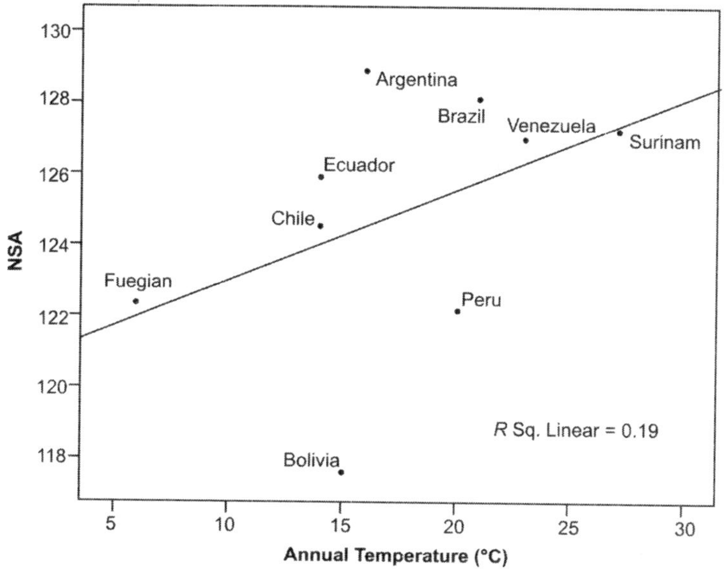

Table 17

NSA/climate correlations for South America

Climate Variable/Proxy		*N*	*r*	*Significance (*.05, 1-tailed)*
L	Latitude (°N/S)	9	−.178	.324
A	Annual mean monthly temperature (°C)	9	+.436	.120
S	Summer (July) mean temperature (°C)	9	+.521	.075
W	Winter (January) mean temperature (°C)	9	+.348	.180

The two Caribbean samples (Dominican Republic and Guadeloupe) were not included in either the North or South American continental analyses given their ambiguous location: most of the region is now linked geopolitically more to North than South America (and so these samples are grouped with North America in the tables), yet the indigenous inhabitants of these two countries (such as the Taínos and the Carib or Kalinago peoples) are linked ethnographically more closely to the northern parts of South America. The Guadeloupe sample is excluded, however, because the four femora are from African black immigrants rather than Native Americans. Correlation results for South America including the Dominican Republic as the sole Caribbean sample are shown in Table 18; correlations are lower, attributable to a low NSA (relative to a warm climate) for the Dominican Republic (NSA 124.3°, $n = 24$).

Table 18

NSA/climate correlations for South America including Caribbean sample

	Climate Variable/Proxy	*N*	*r*	*Significance (*.05, 1-tailed)*
L	Latitude (°N/S)	10	–.175	.315
A	Annual mean monthly temperature (°C)	10	+.371	.146
S	Summer (July) mean temperature (°C)	10	+.454	.094
W	Winter (January) mean temperature (°C)	10	+.296	.203

Regional samples

In addition to the continental analyses, NSA samples from two countries (the USA and France) can be analysed more closely in relation to climatic indices as they each comprise a number of discrete samples provenanced to specific areas or localities (12 for the USA and 18 for France). All the USA samples are Native American; most derive from the SI, with one each from the AMNH, FLMNH, and NHM, while virtually all the French samples derive from the exceptionally well-provenanced collections at the MdlH. Hawaii is excluded from the "continental" USA analyses, being assigned (as a Polynesian sample) to the Pacific region.

USA

Descriptive statistics for the USA samples are given in Table 19. Their locations are shown in Figure 18 and the matched climatic data in Table 20. These regional USA samples provide reasonable geographical coverage and, with the inclusion of the Alaskan (Aleutian Islands) samples, extend across a wide range of climatic zones, spanning almost 20°C in terms of mean annual temperatures.

Figure 18

Locations of the 12 samples from the USA

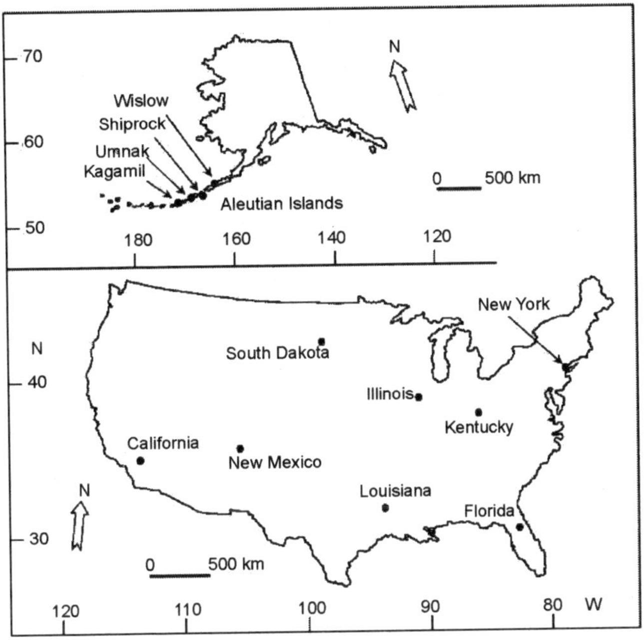

Table 19

Descriptive statistics for the USA samples

State	*Sample*	*N*	*Mean*	*Minimum*	*Maximum*	*SD*
Alaska	Kagamil Is.	129	120.4	109	131	4.32
Alaska	Shiprock Is.	72	118.3	111	126	3.66
Alaska	Umnak Is.	89	121.9	112	128	3.50
Alaska	Wislow Is.	34	119.2	109	126	3.81
California	Chumash	78	126.4	116	136	3.90
Florida	(various)	306	128.9	110	142	5.06
Illinois	Woodland	100	125.3	116	135	4.16
Kentucky	Archaic	129	123.6	113	132	3.97
Louisiana	Mississippian	118	126.6	116	137	4.50
New Mexico	Hawiku	142	123.4	112	133	4.16
New York	Manhattan Is.	13	125.3	117	137	5.02
South Dakota	Arikara	101	120.2	109	131	4.54

Table 20

Climatic data for the USA samples

State/Sample	*Institution*	*NSA*	*Weather Station*	*L*	*A*	*S*	*W*
Alaska/Kagamil Is.	Smithsonian	120.4	*Nikolski*	53	3	9	–1
Alaska/Shiprock Is.	Smithsonian	118.3	*Adak Island*	52	5	10	0
Alaska/Umnak Is.	Smithsonian	121.9	*Nilolski*	53	3	9	–1
Alaska/Wislow Is.	Smithsonian	119.2	*Dutch Harbor*	54	5	11	0
California/Chumash	NHM	126.4	*Los Angeles*	34	18	22	13
Florida/(various)	FLMNH	128.9	*Gainesville*	30	20	27	12
Illinois/Woodland	Smithsonian	125.3	*Carlinville*	39	12	24	–2
Kentucky/Archaic	Smithsonian	123.6	*Louisville*	38	13	25	0
Louisiana/Mississippian	Smithsonian	126.6	*Alexandria*	31	19	27	9
New Mexico/Hawiku	Smithsonian	123.4	*Albuquerque*	35	13	26	1
New York/Manhattan Is.	AMNH	125.3	*New York*	41	12	25	0
South Dakota/Arikara	Smithsonian	120.2	*Pierre*	44	8	24	–7

Figure 19

Scatterplot of NSA and annual mean temperature for the USA groups

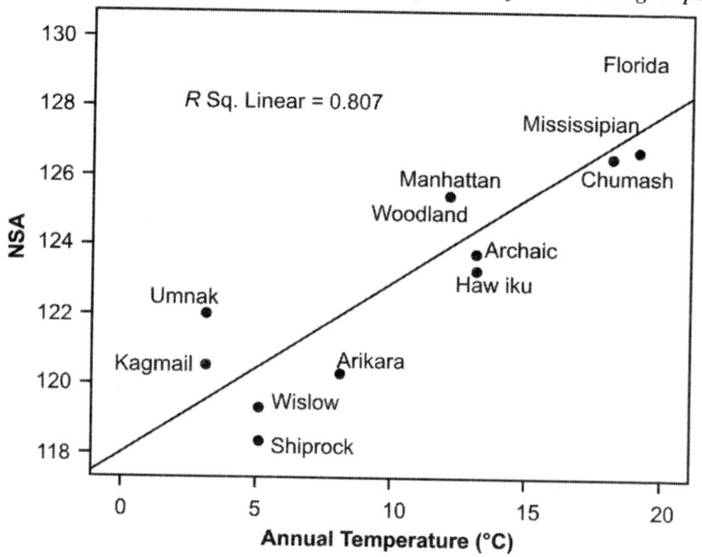

Correlation results are shown in Table 21 and the group distributions for mean annual temperature are plotted in Figure 19. All the correlations between NSA and the climatic indices are highly significant (p 0.005) and follow the anticipated thermal trends.

Table 21

NSA/climate correlation results for the USA

Climate Variable/Proxy		*N*	*r*	*Significance (**.005, 1-tailed)*
L	Latitude (°N/S)	12	–.881	.000**
A	Annual mean monthly temperature (°C)	12	+.899	.000**
S	Summer (July) mean temperature (°C)	12	+.751	.002**
W	Winter (January) mean temperature (°C)	12	+.721	.004**

France

The locations of the 18 samples from France are shown in Figure 20, and the NSA group means and matching weather stations with associated meteorological data are listed in Table 22.

Table 22

NSA means and climatic data for the samples from France

ZONE	*Department*	*Region(s)*	*N*	*NSA*	*Weather Station*	*L*	*A*	*S*	*W*
NORTH	Pas-de-Calais	Nord-Pas-de-Calais	25	126.6	Lille	51	10	17	3
	Aisne	Picardie	8	127.5	Laon	50	9	17	1
	Oise	Picardie	84	126.4	Beauvais	49	9	17	2
	Eure	Haute-Normandie	16	125.6	Rouen	49	10	17	3
	Marne	Champagne-Ardenne	316	126.9	Reims	49	10	18	1
	Paris (Seine)	Île-de-France	102	127.8	Paris	49	11	19	3
	Morbihan	Bretagne	29	124.9	Lorient	48	11	17	6
CENTRAL	Côte-d'Or	Bourgogne	10	128.7	Dijon	47	10	19	2
	Nièvre	Bourgogne	122	127.5	Bourges	47	10	19	3
	Ain	Rhône-Alpes	4	127.0	Lyon	46	11	21	3

Cont'd...

...Cont'd

	Charente	Poitou-Charentes	7	126.1	Poitiers	47	11	18	3
	Savoie	Rhône-Alpes	4	125.8	Valence	45	12	22	3
	Corrèze/ (Auvergne)	Limousin/ Auvergne	4	123.0	Clermont -Ferrand	46	10	19	1
SOUTH	Gironde	Aquitaine	5	126.4	Bordeaux	45	12	20	6
	Lozère	Languedoc-Roussillon	12	128.7	Millau	44	11	19	3
	Aveyron	Midi-Pyrénées	2	119.5	Toulouse	44	12	21	4
	(Alpes)/Bouches-du-Rhône	Provence-Alpes-Côte d'Azur	7	127.0	Marseille	43	15	23	6
	Ariège	Midi-Pyrénées	11	128.5	Perpignan	43	15	23	8

Correlation results on NSA and the climate indices for the French samples are given in Table 23 and the group distributions are plotted against winter temperature in Figure 21. The results show essentially no correlation between NSA and the indices. However, most of the NSA means cluster in a small range (~126°–128°) with a couple of outliers, and span only a limited geographical and climatic range (8° of latitude and a 6° difference in mean annual temperatures), compromising the statistical power of the analyses.

Figure 20

Map of France showing locations of the 18 regional samples

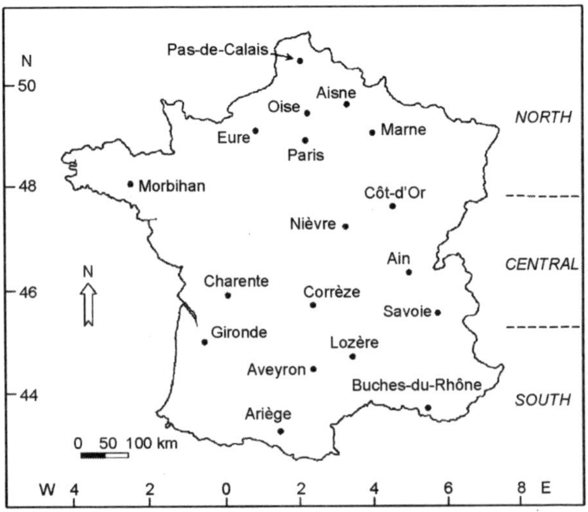

Table 23

NSA/climate correlations for the samples from France

	Climate Variable/Proxy	*N*	*r*	*Significance (*.05, 1-tailed)*
L	Latitude (°N/S)	18	+.126	.309
A	Annual mean monthly temperature (°C)	18	+.020	.469
S	Summer (July) mean temperature (°C)	18	–.020	.469
W	Winter (January) mean temperature (°C)	18	+.037	.442

Figure 21

Scatterplot of NSA and winter temperature for the samples from France

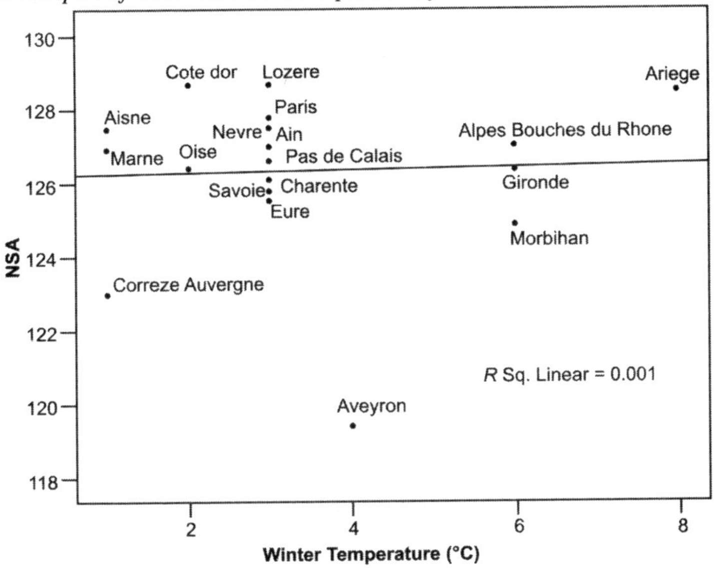

France: Three Zones

One method of exploring these data further is to group the 18 French samples into three zones – north, central, and south – which may emphasize any overall trend in the NSA means across the country. Mean NSA values for these three zones and the matched weather stations (with associated climatic data) are given in Table 24.

Table 24

Mean NSA and climatic data for the three French zones

Zone	*n*	*NSA*	*Weather Station*	*L*	*A*	*S*	*W*
North	580	126.9	Reims	49	10	18	1
Central	15	127.3	Lyon	46	11	21	3
South	37	127.5	Marseille	43	15	23	6

Correlations between NSA and climatic indices for the French samples grouped into three zones are strong (Table 25), although only one (summer temperature) reaches statistical significance by virtue of the minimal number of groups (three) in the analyses. Despite the small variation in NSA across the total French sample of 768 femora from 18 locations, geographical (and climatic) trends nonetheless emerge at this fine level of resolution when the samples are grouped in a manner that minimizes the impact of outliers and emphasizes any overall patterning in the data (Figure 22).

Figure 22

Scatterplot of summer temperature and mean NSA for three French zones

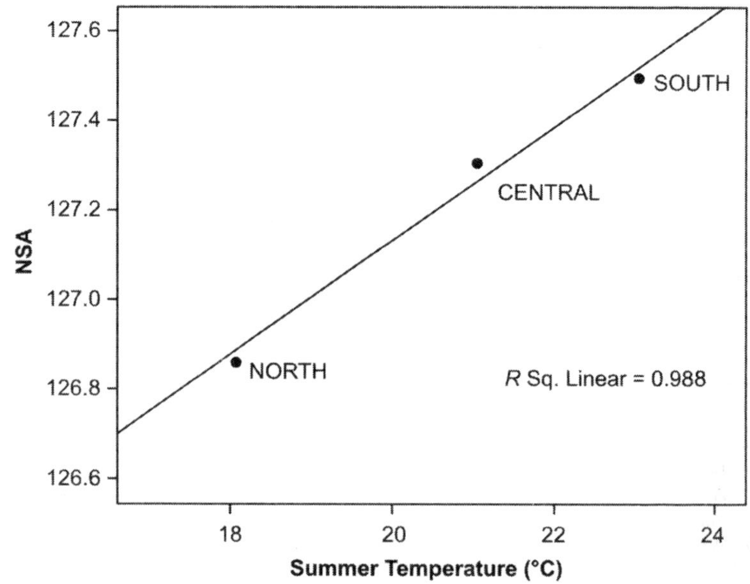

Table 25

NSA/climate correlations for three zones in France

	Climate Variable/Proxy	*N*	*r*	*Significance (*.05, 1-tailed)*
L	Latitude (°N/S)	3	–.975	.072
A	Annual mean monthly temperature (°C)	3	+.848	.178
S	Summer (July) mean temperature (°C)	3	+.994	.035*
W	Winter (January) mean temperature (°C)	3	+.943	.108

The small variation (1°) in mean NSA between the three zones spans barely 10° of latitude, with variation in mean annual temperature of 5°C. Yet, with many of the samples deriving from archaeological (mainly neolithic, *c.* 2500 BP) contexts, and the Paris (urban) sample from the early Christian-era Saint-Marcel cemetery, the local populations may have been relatively stable in the respective zones over considerable time scales. The three zones correspond broadly to distinct climatic zones in France, which traverse the country from the northeast to the southeast, with the north being predominantly a cool temperate maritime zone, the central zone more continental, and the southern zone influenced by the mild Mediterranean climate.

The Americas combined

On ethnographic grounds, the two American continental analyses can be combined into a single Native American analysis. The regional USA samples can also be utilized in this analysis, creating a total of 25 group samples spanning both continents, from the Inuit samples in the north to the Ecuador and Caribbean samples near the equator (excluding the Guadeloupe sample), and with the Fuegians at the southern extreme. Climatic correlations for NSA and results of significance tests for the combined Americas are given in Table 26.

Table 26

NSA/climate correlations for the Americas combined

	Climate Variable/Proxy	*N*	*r*	*Significance (*.05, 1-tailed)*
L	Latitude (°N/S)	25	–.309	.066
A	Annual mean monthly temperature (°C)	25	+.438	.014*
S	Summer (July) mean temperature (°C)	25	+.408	.022*
W	Winter (January) mean temperature (°C)	25	+.333	.052

In contrast to results for the separate North and South American analyses, and even with the anomalous Greenland sample included, two of the climatic correlations now reach statistical significance (at the $p <$ .05 level). The distribution of Native American NSA group means on the annual temperature variable is shown in Figure 23; the r^2 value indicates that 19.2 per cent of NSA variance among the Native American groups is "explained" by mean annual temperature variation. With the exception of the Greenland outlier, the group NSA means are distributed in a reasonably predictable fashion based on thermal expectations. Stronger climatic correlations emerge when the Americas are analysed without the Greenland sample (Table 27 and Figure 24).

Table 27

NSA/climate correlations for the Americas excluding Greenland

	Climate Variable/Proxy	*N*	*r*	*Significance (*.05, 1-tailed **.005, 1-tailed)*
L	Latitude (°N/S)	24	–.480	.009*
A	Annual mean monthly temperature (°C)	24	+.643	.000**
S	Summer (July) mean temperature (°C)	24	+.603	.001**
W	Winter (January) mean temperature (°C)	24	+.473	.010*

Figure 23

NSA and annual temperature in the Americas

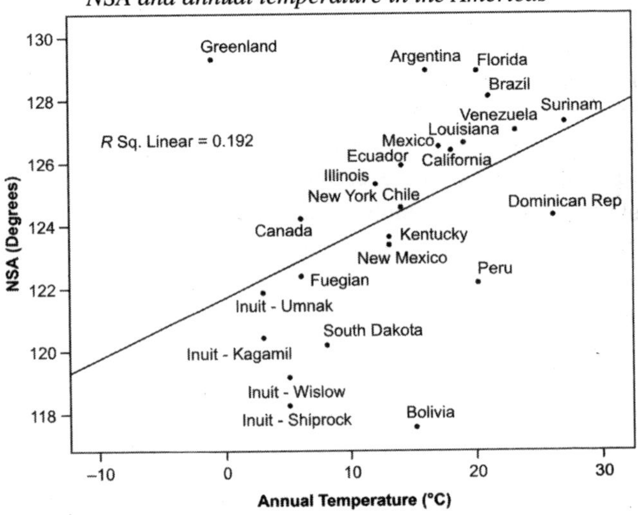

Figure 24

NSA and annual temperature in the Americas excluding Greenland

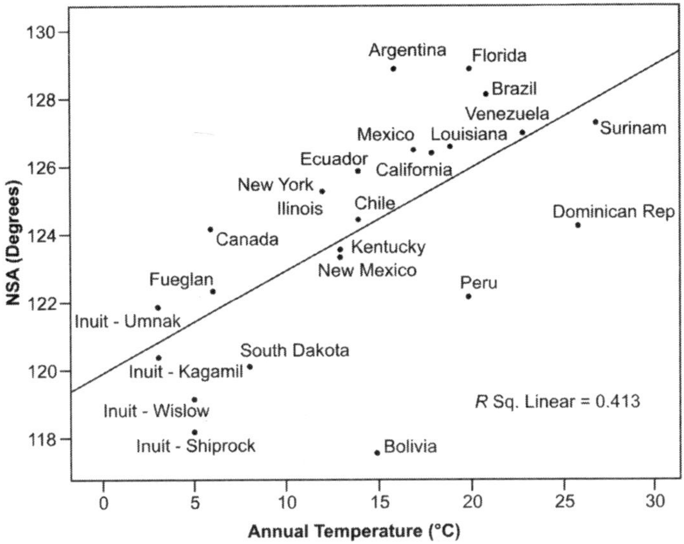

Discussion

The findings shown here on the femoral NSA in modern humans confirm the high magnitude of its variation at the global and group levels, underlining the need to ensure both a wide geographic spread of samples and inclusion of adequate sample sizes before making generalizations as to any patterning or trends that may exist in its variation. Within the larger group samples, the NSA ranges up to approximately 30°, a figure which far exceeds any likely mean differences at a population level. For example, in the largest country sample (England, *n* = 1,271), NSA varies from a minimum of 108° to a maximum of 142°, a range of 32°; similarly, the USA samples (*n* = 987, excluding Alaska and Hawaii) have a 33° range, the French samples (*n* = 768) a range of 30°, and likewise the Canary Islands (*n* = 223, range 30°). Within-group variation is not uncommonly between around 20° and 30° even in smaller samples, e.g., Madagascar (*n* = 62, range 22°), China (*n* = 115, range 26°), Sudan (*n* = 130, range 25°), the Philippines (*n* = 92, range 28°), Vietnam (*n* = 16, range 25°), and Chile (*n* = 49, range 31°).

This extensive NSA database is unique in terms of its total size, extent of group sampling, and geographical coverage. It also has the benefit of the data deriving from measurements taken by a single observer using a hand-held goniometer on the femora themselves, minimizing the confounding effect of femoral anteversion in determining NSA – and avoiding the perils of pooling data from other published sources where techniques may differ substantially (and, in many cases, where the measurements have been taken on radiographs).

In terms of geographical and climatic patterning evident in the data, the results presented here are limited to a preliminary overview at the global and continental levels using comparatively rudimentary statistical analysis to ascertain the likelihood of large-scale trends in mean NSA values. The findings certainly suggest that, notwithstanding the high variability within groups and the small sample sizes in many samples (36 – approximately one-third – of the 101 groups containing fewer than 10 femora), the breadth of sampling reveals an overall significant trend of lower NSA in cooler environments among modern human groups.

A preliminary analysis of the full NSA database (n = 8,271), which comprises 101 samples from around the world, found consistent climatic trends in the NSA (Gilligan, in press). Of the four climatic indices (latitude, mean annual, summer, and winter), the strongest correlations occurred with mean winter temperature, reflecting the role of cold stress in human morphological adaptation to climatic variation. In the present study, climatic trends in NSA are found both between and within the continental regions, with the Pacific region having the highest "continental" mean NSA (129.9°) followed by Africa (128.5°). Climatic trends at the continental level are most marked in Asia but are found on every continent, with the Americas having relatively low NSA group means, a consequence presumably of selection pressure for cold adaptation (and hence a more stocky body shape, and lower NSA) among the likely ancestors of Native Americans in north-eastern Eurasia during the late Pleistocene.

The database lends itself to further analytic exploration with more sophisticated methods which may allow other questions to be addressed and enable more complex relationships to be disentangled. Nonetheless, geographical and climatic patterning exists in the NSA at regional and continental levels in addition to the global trend that emerged in the preliminary analysis. Within Africa, not only are the mean NSA values generally higher than for other continents, most of the higher NSA means are found within the tropics (e.g., Chad 132.3°, Ethiopia 132.2°, Gabon 130.0, Gambia 144.0° – the highest value here, but a very small sample, $n = 2$, Guinea 133.5°, Mali 130.8°, Senegal 132.5°, and Sudan 130.5°), while most of the lower mean NSA values occur in cooler parts of the continent (e.g., Algeria 123.4°, Morocco 120.0°, Mozambique 120.8°). The regional trends in eastern Asia, the Pacific, and North and South America appear less prominent but are similarly suggestive. Access to skeletal samples from Australia has been limited owing to cultural sensitivities; all that can be said is that the single mean NSA of 130.2° from South Australia (the Roonka burial ground, $n = 47$) is in the higher range on a global scale, as might be expected, and is comparable with the only previously published mean NSA for Australian Aborigines (Davivongs, 1963) of 127.8°.

Of special anthropological interest are the NSA results for the 11 "other groups" where the data are presented separately from countries and territories. Among these, the higher NSA values tend again to occur in cooler environments (e.g., Inuit – in this instance, the term Eskimo or Aleut might be less inaccurate, as the samples derive from the Aleutian archipelago in Alaska – 120.2°, Fuegian 122.4°, Okhotsk 124.8°, and the Ainu and Sami 125.3°), while higher NSA values generally occur among those groups inhabiting warmer regions (e.g., Andaman Islanders 136.1°, Veddah 132.3°, central African Pygmies 131.6°, and Philippine Negritos 130.9°).

There are obvious limitations in the present analyses, some of which may be improved to further enhance the usefulness of this NSA database. Firstly, these results have utilized relatively simple climatic indices – latitude, mean annual temperature, and mean monthly summer and winter temperatures. Surprisingly, given that latitude is at best a crude approximation for average thermal conditions, global and continental NSA trends have emerged on this variable. Nevertheless, more sophisticated meteorological indices – such as average maximum and minimum temperatures and, ideally, the more physiologically meaningful indices of thermal stress such as wind chill and apparent temperature (e.g., Steadman, 1984; Quayle and Steadman, 1998) – would provide superior measures of the local thermal environments.

A major shortcoming, however, is the need to control for some of the non-environmental variables that have been identified in other studies to influence variation in the NSA among human groups. Of these, the most salient relates to habitual activity levels and degree of mobility in different population samples. As discussed above, a number of the earlier studies have attributed a greater role for these activity factors in NSA variability than climate (e.g., Trinkaus, 1993; Anderson and Trinkaus, 1998), although the debate is far from resolved (e.g., Weaver, 2003). With respect to the NSA database presented here, there is evidently scope to investigate this issue by examining the data in relation to additional sample categories, for instance, by dividing the groups according to lifestyle (e.g., forager, agricultural, and urban). Another factor that should be taken into account, given the further aim of exploring whether greater "cultural buffering" from the thermal environment may exert an influence on patterns of morphological variation, is the extent to which the various population groups sampled here employ thermally effective clothing on a regular basis.

Cultural buffering

The significance of cultural buffering – mainly, though not limited to, the habitual use of clothing – is that, as a behavioural adaption to the thermal environment, it might be expected to exert an ameliorating effect on the development of morphological adaptations, and vice versa (e.g., Gilligan, 2007b). Such interactions between biological and behavioural adaptations may affect the extent to which clothing is required as thermal protection, among other repercussions (e.g., Gilligan, 2007a, 2007c). Furthermore, the regular use of clothing may also act as a confounding variable in assessing the possible role of mobility on the NSA. The reason this may be the case is that the lifestyle categories commonly used to classify levels of habitual activity – forager, agricultural, and urban (e.g., Anderson and Trinkaus, 1998: 284) – may themselves be associated systematically with differing levels of clothing use. Foragers such as the Australian Aborigines typically wore no clothing, among whom its use appears to relate exclusively to minimal thermal requirements (Gilligan, 2008). The Andaman Islanders and Fuegians are other examples, while the advent of agricultural practices – in terms of the evidence from prehistoric archaeology – is associated with a marked focus on the production of clothing, especially thermally efficient textile garments (Gilligan, 2007d). Similar trends in patterns of clothing use are apparent among early Holocene foragers who adopted a more sedentary lifestyle, with or without agricultural, horticultural, or pastoralist activities, while those communities that developed a fully urban lifestyle were likewise characterized by the regular use of clothing. In other words, the categories that have been employed to distinguish differing levels of mobility or habitual activity may also distinguish differing levels of habitual clothing use, which might create a thermal basis for any apparent trends in the NSA between these groups.

Conclusions

Despite its limitations, the data on the NSA among human groups and the environmental analyses presented here document considerable variation in this morphological parameter both within and between groups, variation which nonetheless manifests an overall association with geographical variation in average temperature regimes. Significant and consistent trends in climatic patterning of the NSA at a population level are apparent at both the global and continental scales of analysis. To this extent, the NSA may vary in concert with other aspects of human morphology – notably body shape – which show environmental patterning as a reflection of thermally adaptive influences. Further analyses and refinements in methodology may be expected to clarify these findings and address some of the unresolved questions that surround the NSA, which may shed light on hitherto unexplored aspects of the interaction between biological and behavioural adaptations among modern humans.

Acknowledgements

The author thanks David Bulbeck, Colin Groves, Marc Oxenham, Colin Pardoe, and Peter White for their assistance at various stages throughout the course of this research, and Erik Trinkaus for his useful advice on measuring femoral neck-shaft angles. I also wish to express my gratitude to the curators, managers, and other staff who facilitated access to the collections (listed alphabetically by institution): Gisselle Garcia (American Museum of Natural History, New York), Pasuk Mahakkanukrauh (Faculty of Medicine, Chiang Mai University), Supichya Chandraphak (Congdon Anatomical Museum, Bangkok), Donna Ruhl (Florida Museum of Natural History), Henry de Lumley, Amélie Vialet, and Stéphanie Renault (Institut de Paléontologie Humaine, Paris), Mercedes Okumura (Leverhulme Centre for Human Evolutionary Studies, Cambridge), Philippe Mennecier and others – notably Aurélie, Véronique and Liliana – at the Museé de l'Homme (Paris), Reiko Kono (National Museum of Nature and Science, Tokyo), Robert Kruszynski (Natural History Museum, London), Olivia Herschensohn (Peabody Museum, Harvard University), Hirofumi Matsumura (Sapporo Medical University), David R. Hunt (Smithsonian Institution, Washington), Keryn Walshe and Rebekah Candy (South Australian Museum), and Eric Cook (Indigenous representative for Roonka). This work was supported by an ANU University House Postgraduate Research Scholarship and a Fieldwork Grant from the Faculty of Arts, Australian National University.

References

Allen, J.A. 1877. The influence of physical conditions in the genesis of species. *Radical Review* 1: 108–140.

Anderson, J.Y. and Trinkaus, E. 1998. Patterns of sexual, bilateral and interpopulational variation in human femoral neck-shaft angles. *Journal of Anatomy* 192: 279–285.

Arensburg, B. 1994. Femoral neck-shaft angle: a response to Trinkaus. *Journal of Human Evolution* 27: 449–450.

Ashton, K.J., Tracy, M.C., and de Queiroz, A. 2000. Is Bergmann's rule valid for mammals? *American Naturalist* 156: 390–415.

Bayle, P., Macchiarelli, R., Trinkaus, E., Duarte, C., Mazurier, A., and Zilhão, J. 2010. Dental maturational sequence and dental tissue proportions in the early upper paleolithic child from Abrigo do Lagar Velho, Portugal. *Proceedings of the National Academy of Sciences* 107: 1338–1342.

Beals, K.L., Smith, C.L., and Dodd, S.M. 1983. Climate and the evolution of brachycephalization. *American Journal of Physical Anthropology* 62: 425–437.

Bergmann, C. 1847. Uber die verhältniesse der wärmeokonomie der thiere zu ihrer grösse. *Gottingen Studien* 1: 595–708.

Cheng, X.G., Lowett, G., Boonen, S., Nicholson, P.H.F., Brys, P., Nijs, J., and Dequeker, J. 1997. Assessment of the strength of proximal femur in vitro: relationship to femoral bone mineral density and femoral geometry. *Bone* 20: 213–218.

Collier, S. 1989. The influence of economic behaviour and environment upon robusticity of the post-cranial skeleton: a comparison of Australian Aborigines and other populations. *Archaeology in Oceania* 24: 17–30.

Crognier, E. 1981. Climate and anthropometric variations in Europe and the Mediterranean area. *Annals of Human Biology* 8: 99–107.

Davivongs, V. 1963. The femur of the Australian Aborigine. *American Journal of Physical Anthropology* 21: 457–467.

Doherty, M., Courtney, P., Doherty, S., Jenkins, W., Maciewicz, R.A., Muir, K., and Zhang, W. 2008. Nonsperical femoral head shape (pistol grip deformity), neck shaft angle, and risk of hip osteoarthritis. *Arthritis & Rheumatism* 58: 3172–3182.

Geist, V. 1987. Bergmann's rule is invalid. *Canadian Journal of Zoology* 65: 1035–1038.

Gilligan, I. 2007a. Neanderthal extinction and modern human behaviour: the role of climate change and clothing. *World Archaeology* 39: 499–514.

Gilligan, I. 2007b. *Another Tasmanian Paradox: Clothing and Thermal Adaptations in Aboriginal Australia*. BAR International Series. Oxford: Archaeopress.

Gilligan, I. 2007c. Clothing and modern human behaviour: prehistoric Tasmania as a case study. *Archaeology in Oceania* 42: 102–111.

Gilligan, I. 2007d. Clothing and farming origins: the Indo-Pacific evidence. *Bulletin of the Indo-Pacific Prehistory Association* 27: 10–24.

Gilligan, I. 2008. Clothing and climate in Aboriginal Australia. *Current Anthropology* 49: 487–495.

Gilligan, I. 2010a. Femoral neck-shaft angle and climate: preliminary report on a global study. In K.E. Weiss (ed.), *Trends in Physical Anthropology*. Hauppauge, NY: Nova Science.

Gilligan, I. 2010b. The prehistoric development of clothing: archaeological implications of a thermal model. *Journal of Archaeological Theory and Method* 17: 15–80.

Gilligan, I. and Bulbeck, D. 2007. Environment and morphology in Australian Aborigines: a re-analysis of the Birdsell database. *American Journal of Physical Anthropology* 134: 75–91.

Gnudi, S., Sitta, E., and Fiumi, N. 2007. Bone density and geometry in assessing hip fracture risk in post-menopausal women. *British Journal of Radiology* 80: 893–897.

Gould, S.J. 1966. Allometry and size in ontogeny and phylogeny. *Biological Review* 41: 587–640.

Hiernaux, J. 1963. Heredity and environment: their influence on human morphology. A comparison of two independent lines of study. *American Journal of Physical Anthropology* 21: 575–589.

Holliday, T.W. 1997. Body proportions in late Pleistocene Europe and modern human origins. *Journal of Human Evolution* 32: 423–447.

Holliday, T.W. 1999. Brachial and crural indices of European late upper palaeolithic and mesolithic humans. *Journal of Human Evolution* 36: 549–566.

Holliday, T.W. and Falsetti, A.B. 1995. Lower limb length of European early modern humans in relation to mobility and climate. *Journal of Human Evolution* 29: 141–153.

Holt, B.M. 2003. Mobility in upper paleolithic and mesolithic Europe: evidence from the lower limb. *American Journal of Physical Anthropology* 122: 200–215.

Houghton, P. 1990. The adaptive significance of Polynesian body form. *Annals of Human Biology* 17: 19–32.

Igbigbi, P.S. 2003. Collo-diaphysial angle of the femur in East African subjects. *Clinical Anatomy* 16: 416–419.

Katzmarzyk, P.T. and Leonard, W.R. 1998. Climatic influences on human body size and proportions: ecological adaptations and secular trends. *American Journal of Physical Anthropology* 106: 483–503.

Mayr, E. 1956. Geographical character gradients and climatic adaptation. *Evolution* 10: 105–108.

Meiri, S. and Dayan, T. 2003. On the validity of Bergmann's rule. *Journal of Biogeography* 30: 331–351.

O'Connell, J.F. and Allen, J. 2004. Dating the colonization of Sahul (Pleistocene Australia – New Guinea): a review of recent research. *Journal of Archaeological Science* 31: 835–853.

Ogata, K. and Goldsand, E.M. 1979. A simple biplanar method of measuring femoral anteversion and neck-shaft angle. *Journal of Bone and Joint Surgery* 61: 846–851.

Olsen, M., Davis, E.T., Gallie, P.A.M., Waddell, J.P., and Schemitsch, E.H. 2009. The reliability of radiographic assessment of femoral neck-shaft and implant angulation in hip resurfacing arthroplasty. *Journal of Arthroplasty* 24: 333–340.

Pearson, O.M. 2000. Activity, climate, and postcranial robusticity. *Current Anthropology* 41: 569–607.

Pearson, O.M. and Millones, M. 2003. Postcranial reflections of climatic adaptation and habitual activity in Tierra del Fuego [abstract]. *American Journal of Physical Anthropology* 120: 166.

Pearson, O.M. and Millones, M. 2005. Rasgos esqueletales de adaptación al clima y a la actividad entre los habitantes aborígenes de Tierra del Fuego [Skeletal features showing adaptation to climate and activity of aboriginal inhabitants of Tierro del Fuego]. *Magellania* 33: 37–50.

Pulkkinen, P., Eckstein, F., Lochmüller, E-M., Kuhn, V., and Jämsä, T. 2006. Association of geometric factors and failure load level with the distribution of cervical vs. trochanteric hip fractures. *Journal of Bone and Mineral Research* 21: 895–901.

Quayle, R.G. and Steadman, R.G. 1998. The Steadman wind chill: an improvement over present scales. *Weather and Forecasting* 13: 1187–1193.

Roberts, D.F. 1978. *Climate and Human Variability*. Chicago, IL: Cummings.

Ruff, C.B. 1991. Climate and body shape in hominid evolution. *Journal of Human Evolution* 21: 81–105.

Ruff, C.B. 1993. Climatic adaptation and hominid evolution: the thermoregulatory imperative. *Evolutionary Anthropology* 2: 53–60.

Ruff, C.B. 1994. Morphological adaptation to climate in modern and fossil hominids. *Yearbook of Physical Anthropology* 37: 65–107.

Ruff, C.B. 2002. Variation in human body size and shape. *Annual Review of Anthropology* 31: 211–232.

Ruff, C.B., Trinkaus, E. and Holliday, T.W. 2002. Body proportions and size. In J. Zilhão and E. Trinkaus (eds), *Portrait of the Artist as a Child: the Gravettian Human Skeleton from the Abrigo do Lagar Velho and its Archaeological Context*. pp. 365–391. Lisbon: Instituto Portugẽs de Arqueologia.

Ruff, C.B., Holt, B.M., Sládek, V., Berner, M., Murphy, W.A., zur Nedden, D., Seidler, H., Recheis, W. 2006. Body size, body proportions, and mobility in the Tyrolean "Iceman". *Journal of Human Evolution*. 51: 91-101.

Schreider, E. 1975. Morphological variations and climatic differences. *Journal of Human Evolution* 4: 529–539.

Sládek, V., Berner, M., and Sailer, R. 2006. Mobility in central European late Eneolithic and early Bronze Age: femoral cross-sectional geometry. *American Journal of Physical Anthropology* 130: 320–332.

Spindler, K. 2004. *The man in the Ice: the Preserved Body of a Neolithic Man Reveals the Secrets of the Stone Age*. London: Weidenfeld and Nicolson.

SPSS Inc. 2007. *SPSS Graduate Pack 16.0 for Windows*. Chicago, IL.

Steadman, R.G. 1984. A universal scale of apparent temperature. *Journal of Applied Meteorology* 23: 1674–1687.

Stock, J.T. 2006. Hunter-gatherer postcranial robusticity relative to patterns of mobility, climatic adaptation, and selection for tissue economy. *American Journal of Physical Anthropology* 131: 194–204.

Trinkaus, E. 1981. Neanderthal limb proportions and cold adaptation. In C.B. Stringer (ed.), *Aspects of Human Evolution*. pp. 187–223. London: Taylor & Francis.

Trinkaus, E. 1993. Femoral neck-shaft angles of the Qafzeh-Skhul early modern humans, and activity levels among immature Near Eatern Middle Paleolithic hominids. *Journal of Human Evolution* 25: 393–416.

Trinkaus, E. 1994. Femoral neck-shaft angles: a reply to Arensburg. *Journal of Human Evolution* 27: 451–455.

Trinkaus, E., Churchill, S.E., Ruff, C.B., and Vandermeersch, B. 1999. Long bone shaft robusticity and body proportions of the Saint-Césaire 1 Chṛtelperronian Neanderthal. *Journal of Archaeological Science* 26: 753–773.

Trinkaus, E. and Ruff, C.B. 1999. Diaphyseal cross-sectional geometry of Near Eastern middle palaeolithic humans: the femur. *Journal of Archaeological Science* 26: 409–424.

Trinkaus, E., Stringer, C.B., Ruff, C.B., Hennessy, R.J., Roberts, M.B., and Parfitt, S.A. 1999. Diaphyseal cross-sectional geometry of the Boxgrove 1 Middle Pleistocene human tibia. *Journal of Human Evolution* 37: 1–25.

Walensky, N.A. and O'Brien, M.P. 1968. Anatomical factors relative to the racial selectivity of femoral neck fracture. *American Journal of Physical Anthropology* 28: 93–96.

Weatherbase[SM] (www.weatherbase.com). Great Falls, Virginia: Canty and Associates.

Weaver, T.D. 2003. The shape of the Neandertal femur is primarily the consequence of a hyperpolar body form. *Proceedings of the National Academy of Sciences* 100: 6926–6929.

Weaver, T.D. and Steudel-Numbers, K. 2005. Does climate of mobility explain the differences in body proportions between Neandertals and their upper paleolithic successors? *Evolutionary Anthropology* 14: 218–223.

Weidenreich, F. 1945. The brachycephalization of recent mankind. *Southwestern Journal of Anthropology* 1: 1–54.

Weinstein, K.J. 2005. Body proportions in ancient Andeans from high and low altitudes. *American Journal of Physical Anthropology* 128: 569–585.

Wescott, D.J. 2006. Effect of mobility on femur midshaft external shape and robusticity. *American Journal of Physical Anthropology* 130: 201–213.

Wescott, D.J. and Cunningham, D.L. 2006. Temporal changes in Arikara humeral and femoral cross-sectional geometry associated with horticultural intensification. *Journal of Archaeological Science* 33: 1022–1036.

Zilhão, J. and Trinkaus, E. 2002. A brief history. In J. Zilhão and E. Trinkaus (eds), *Portrait of the Artist as a Child: the Gravettian Human Skeleton from the Abrigo do Lagar Velho and its Archaeological Context*. pp. 13–27. Lisbon: Instituto Portugês de Arqueologia.

3 Endocast of Saldanha Skull

Revised Description, Comparative Status and Behavioural Possibilities

Charles M. Musiba, Carol Laciny and Cassian C. Magori

Continued discoveries of hominin remains representing the genus *Homo* in sub-Saharan Africa and Europe complicate the evolutionary picture of modern humans (Asfaw et al., 2002: 318–319; Gabunia et al., 2001: 129; Haile-Selassie et al., 2004; Martinez and Arsuaga, 1997: 284). The origin of anatomically modern (a.m.) humans has recently gained renewed attention by palaeoanthropologists (e.g., Bräuer, 1984: 328; Bräuer et al., 2003; Clark et al., 2003: 748–749; Grün and Beaumont, 2001: 477–479; Grün et al., 1996: 500; Rightmire, 1998: 218; Stringer, 1992: 217, 219; Stynder et al., 2001: 372, 378; White et al., 2003: 742). In considerations of the transition from "archaic" to "anatomically modern" (a.m.) *Homo sapiens*, increasing evidence from sub-Saharan Africa demands special considerations. Numerous sites in southern Africa such as Border Cave (Butzer et al., 1978: 336; Grün and Beaumont, 2001: 477),

Hoedjiespunt 1 at Saldanha Bay (Stynder et al., 2001: 372); Florisbad (Grün et al., 1996: 500), and Klasies River Mouth (Singer and Wymer, 1982: 205; Thackeray, 2002: 752) have produced remains of a.m. *Homo sapiens* with probable ages ranging from 259 to 100 Ka. Based on the antiquity of fossil hominins from these sites, it is possible, therefore, that anatomically modern humans appeared both in southern and the horn of Africa earlier than anywhere else in the world (Bräuer et al., 2003; Klein, 1983: 25; Rightmire and Deacon, 2001: 538; Sillen and Morris, 1996: 503).

Although phylogenetic relationships of hominin remains representing the genus *Homo* are much more complex and mosaic in nature (e.g., Conroy, 2002; Eldredge and Tattersall, 1982: 156; Hofman, 1983: 102; Wood and Richmond, 2000: 20), *Homo heidelbergensis* appears to represent a "transitional" population of archaic *Homo sapiens* that inhabited sub-Saharan Africa during the late Middle and early Upper Pleistocene (ranging from ~ 250–100 Ka; Asfaw et al., 2002: 318–319; Howell, 1978: 162). Many of the remains of archaic *Homo sapiens* in Africa such as Ndutu (Clark, 1976, 1990; Mturi, 1976), Elandsfontein (Saldanha), Kabwe (Clark et al., 1950; Seidler et al., 1997; Broken Hill: Smith, 1928), and Bodo (Adefris, 1992) which were initially included in the *Homo sapiens rhodesiensis* hypodigm, have been reassigned to *Homo heidelbergensis* (a broader group including the European archaic *Homo sapiens*; Bräuer, 1984, 1992; Bräuer and Stringer, 1997). In this paper, we present a morphological review of the original endocast of the re-assembled Saldanha specimen, especially features that remain most tentative (observable circular deformation within the parietal region of the right hemisphere here attributed to possible meningioma) and overall brain volume.

Materials

The Saldanha skull

The Saldanha calotte from which the endocast used in this study was cast in plaster, was re-assembled from 27 cranial fragments to form the virtually (almost) complete frontal and parietal regions, the *planum occipitale* and upper part of the *planum nuchale* of the occipital bone, and the superior and posterior portion of the right temporal region.

Lacking are the left temporal, most of the right temporal, the basi- and exoccipital regions, and the remainder of the cranial base. The posterior aspect of the reconstructed calotte (Figure 1: Drennan, 1953a: 792) has a rather asymmetric appearance: a more medially depressed portion of the right *planum nuchale* appears to have been misaligned during reconstruction (Singer, 1954: 456) reportedly for the sake of emphasizing the "primitive" character of the back of the skull of the Saldanha individual (see below).

Figure 1

Lateral view of the reconstructed Saldanha skull showing the overall cranial morphology

Compared to the Kabwe skull, the Saldanha calotte shows no distinction in respect of the slope of the forehead (Drennan, 1953a: 627). Supraorbital tori of the two specimens are similar in lateral profile, although the Saldanha individual is clearly more gracile. Figure 2 shows lateral profiles of the Saldanha calotte and Kabwe skull drawn from their X-rays oriented along the nasion-inion line. Posterior to vertex the profiles overlap, apparently due to more spherical form of the Saldanha calotte's posterior portion compared to that of the anteroposteriorly elongated Kabwe skull. In *norma verticalis*, the two specimens are similarly pear-shaped (Pycraft, 1928: 17; Singer, 1954: Plate 1) and both show their greatest width slightly posterior to the (presumed in the case of Saldanha) auditory aperture. Viewed posteriorly, the sides of the Saldanha calotte rise somewhat more steeply than those of the Kabwe skull. The greatest width of the Saldanha calotte appears to lie higher than that of the Kabwe skull (Drennan, 1955: 628). The brain falls below the lower limit of modern human values for brain size (endocranial capacity here) (Table 2).

Figure 2

Superimposed outlines of lateral profiles of Saldanha and Kabwe crania, after X-rays (actual size) to show the form of the posterior cranial fossa. Kabwe – Saldanha. Scaled size

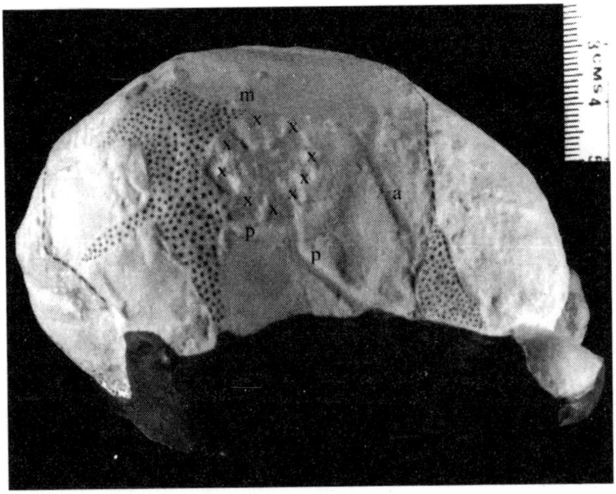

In overall size and shape, the Saldanha calotte is smaller and more spherical than the Kabwe skull. On account of its smaller size, the Saldanha individual has been considered to have been a female ("gynomorphic") (Drennan, 1955: 617; Singer, 1954: 357), but Rightmire (1993) believes that "the Elandsfontein skullcap is so incomplete that no accurate sex determination can be made." The more spherical conformation of the posterior aspect of the Saldanha calotte might be due to a scaling effect of smaller size in the Saldanha individual compared to the Kabwe individual (see Holloway et al., 2004; Weidenreich, 1941: 378).

Table 1

Cranial morphology of transitional sub-Saharan specimens compared with Homo erectus and modern forms

Specimen	*Maximum length (glabella-opistho-cranion) (mm)*	*Maximum breadth*	*Ectocranial index (br/ln)*	*Max. Supra-orbital thickness (mm)*	*Max. occipital thickness (mm)*	*Probable Age (mya)*
Homo erectus	194.5	141	0.72	19	NA	1.2–1.5
Ave. (range)[1]	183–206	132–150				
Saldanha[2]	200	144	0.72	13	12	> 0.125
Kabwe[3]	208.5	144.5	0.69	22	NA	< Saldanha
Ndutu[4]	NA	NA	NA	8.6	NA	0.5–0.6
Eyasi[5]	191	142	0.75	NA	NA	0.130
Bodo[6]	NA	NA	NA	17 – 18	NA	0.12–0.07
Laetoli[7]	205	140	0.68	NA	NA	0.120–0.07
a.m. *H. sapiens*[8]	185.6 (158 – 203)	133.6 (124–157)	0.728 (0.652–0.893)	6.05 (5.8–6.3)	15	0.130–present

NA = not available

Sources: 1. Leakey, 1976: 573; Rightmire, 1979: 100, 103; Walker and Leakey, 1978: 62.
2. Singer, 1954: 351, 354.
3. Pycraft, 1928: 2, 15; Morant, 1928: 341.
4. Mturi, 1976: 485.
5. Mehlman, 1984: 498; Weinert et al., 1939: 263, 264.
6. Conroy et al., 1978: 67.
7. Day et al., 1980: 55.
8. Weidenreich, 1943: 116, 119, and 163.

Table 2

Ranges and means of cranial capacity (cm^3) and endocranial length-breadth indices transitional forms, Homo erectus and a.m. Homo sapiens.

Hominin Grade	*n*	*Average*	*Min.*	*Max.*	*Endocranial Index*	*References*
a.m. *Homo sapiens* (various races)	8	1400.0	1270.0	1610.0	0.843 (0.734– 0.950)	Weidenreich, 1945: 17; 1943: 116
Chinese	32	NA	NA	NA	0.787	Hirschler, 1942: 118
Easter Island	25	1400.5	1132.0	1644.0	NA	Murrill, 1968: Tables 5a and 6b

Cont'd...

...Cont'd

Early Western Asia	6	1545.0	1491.4*	1598.8*	NA	Trinkaus, 1984: 272
Early European	11	1576.8	1305.8*	1598.8*	NA	Trinkaus, 1984: 272
Transitional						
Western Asian Neanderthal	4	1525.3	1129.3*	1921.3*	NA	Trinkaus, 1984: 272
European Neanderthal	6	1509.8	1209.8*	1809.6*	NA	Trinkaus, 1984: 272
European and African Neanderthal	6	1493.0	1370.0	1640.0	0.793	Holloway, 1980: 293; Olivier and Tissier, 1975: 356
Sub-Saharan Africa (Saldanha, Kabwe, Laetoli)	3	1227.0–1242.0	1200.0	1280.0–1325.0		Drennan, 1953a: 793; Singer, 1954: 351; Smith, 1928: 58
Petralona	1	1155.0	NA	NA	> 0.73	Murrill, 1982: 72
Solo	5	1151.0	1090.0	1251.0	0.774	Holloway, 1980: 286, 293
Erectus						
Indonesia	6	942.0	775.0	1127.0	0.735	Holloway, 1980: 293; Olivier and Tissier, 1975: 355
China	5	1043.0	915.0	1225.0	0.765	Weidenreich, 1943: 116; 1945: 16
Africa	2	922.0	850.0	1067.0	0.720	Walker and Leakey, 1978: 62; Rightmire, 1979: 102

(sexes pooled; NA = not available; * = 2 standard deviations from mean). Petralona's index value is based on ectocranial index value (see text).

In summary, the morphological and contextual evidence of sub-Saharan African transitional hominins would indicate that most metrical and general cranial characteristics (Table 1, Figures 1 and 2) lie within the range of a.m. *Homo sapiens*. Superciliary robusticity in transitional forms exceeds that characteristic in modern forms. Transitional cranial fossils in sub-Saharan Africa are associated indirectly with Acheulian artefacts, as well as with yet to be fully described evidence of butchery of large mammals and with Acheulian material in direct association at Bodo D'Ar (Conroy et al., 1978: 67). At

Elandsfontein, several handaxes were located at one site between and around the bones of extinct elephants (Klein et al., 2007). Contextual evidence suggests that this archaic modern morphology characterized early *Homo sapiens* during a period that extended roughly from 100–600 Ka.

The brain morphology of the transitional hominins represented by these sub-Saharan specimens can be studied using endocranial casts. The Kabwe material has been described in detail (Pycraft, 1928), including its endocranial morphology (Smith, op. cit.). Other upper Middle and early Later Pleistocene sub-Saharan cranial evidence is less complete, and until recently endocranial details have seldom been reported (Conroy, 2002; Dreyer, 1936, 1938; Holloway et al., 2004; see also Day et al., 1980; Magori and Day, 1983; Saban, 1983). While the morphology of the Saldanha calotte has been reported (Drennan, 1953a, b; Singer, 1954), its endocranial details briefly described by Holloway et al. (2004) have yet to be described in depth. A study of an original endocast of the reconstructed Saldanha specimen is reported here. Original and cast specimens of the unreconstructed cranial fragments were not available for this study, so some diagnostic features of the endocast are restricted. Features remaining most tentative include occipital *petalia*[1] and overall brain volume.

Geological context of the Saldanha skull

The Elandsfontein hominin occurrence comprises numerous hominin cranial fragments, along with faunal and archaeological remains. The complex geological picture shows periods of aridity interspersed with periods of moister conditions to form tortuous sub-surface ridges of ferricrete (Butzer, 1973: 237; Compton and Franceschini, 2005). The fossiliferous horizon of this depositional context consists of a discrete "sub-horizontal zone of low-grade calcareous concretions" whose variability of thickness and degree of calcification may indicate that more than one horizon is represented in different segments of the site area (op. cit.: 236). The faunal and archaeological materials were found on a depositional surface upon which remains were "quickly buried in sediment on a gently undulating non-dunal surface" (op. cit.: 234). Fossiliferous accumulations are found on the deflation surface of windblown sands in what once may have held a waterhole or vleis (marsh) in an otherwise arid environment (Klein, 1978; Singer and Wymer, 1968: 72). Palynological evidence suggests higher rainfall than at present.

The hominin cranial fragments were found on the surface near or with remains of antelope, hippopotamus, and elephant, as well as extinct species of baboon, giraffids, saber toothed felid, lion, squid, giant buffalo, and other bovids (Hendey, 1969: Table 6; Singer and Crawford, 1957: 17). Partridge has noted that more than 50 per cent of the mammalian fauna are extinct species, and Vrba has reviewed comparisons between bovids at Elandsfontein and those from Olduvai Gorge:

> Among the EF (Elandsfontein) Bovidae *Robaticeras arambourgi* is also known from Olduvai Beds III and IV …, although the EF form is more advanced in some respect. The *Gazella sp.* is possibly conspecific with a species from Olduvai Middle and Upper Bed II … while the EF *Taurotragus* is decidedly more advanced than *Taurotragus arkelli* from Olduvai Bed IV. (Vrba, 1982: 738)

Comparison with fauna from Beds II through IV at Olduvai Gorge suggests dates somewhat younger than the age of Bed IV (greater than 600 Kya; Leakey and Hay, 1982: 759; Manega, 1993; Singer, 1954: 349). Vrba (1982: 739) believes that Saldanha is datable faunally to 300–600 Kya. Fluorine dating suggests contemporaneity between the hominin and some of the animal remains at Elandsfontein (Klein et al., 2007; Singer and Wymer, 1968: 65).

A profusion of Early (Acheulian), Middle, and Late Stone Age artefacts suggests repeated occupation of the Elandsfontein site, beginning perhaps 125–130 Kya (Klein, 1973: 312). The Saldanha cranial fragments have been associated with Acheulian artefacts (Drennan, 1953a: 793). While there is "still" no direct proof of this association (Singer, 1954: 347),

> one of the most persuasive arguments for an association between the main fossil horizon and the Acheulian industry is the fact that, in all South African sites which have yielded artifacts and fauna in association with anatomically modern man, the stone industry has been Middle Stone Age and the fauna has been of recent type (Florisbad

> Faunal Span); the main fossil fauna at Elandsfontein is, in contrast, of decidedly archaic character in conformity with the anatomical characteristics of the hominid specimen .... (Partridge, 1982: 646)

Moreover, the Upper Duricrust lying above the fossiliferous horizon appears to predate the local Middle Stone Age. Furthermore, stratigraphic comparisons with other erosional basins within the area suggest either Acheulian or Middle Stone Age cultural associations that seem to be collaborated biostratigraphically with probable age of 700–400 Kya (Klein and Cruz-Uribe, 1991: 72). Recent faunal and taphonomic analysis of Elandsfontein assemblage based on biostratigraphic correlations by Klein et al. (2007) place the Upper Duricrust layer at the age of ca. 600 Kya consistent with initial biostratigraphic correlated age estimates by Vrba (1982).

Methods

The endocranial aspect of the Saldanha calotte was cast in plaster in Cape Town in 1965, which has produced an overall shrinkage of about four per cent due to life span of material used to produce the endocast. The endocast consists of frontal, parietal, superior occipital surfaces, and a small posterior portion of the right temporal lobe.

Maximum breadth and anterior-posterior length for both hemispheres were obtained using visual and manual inspection and spreading calipers. Greatest breadth on the Saldanha endocast occurs near the posterior extreme of the lambdoid (posterior) branch of the middle meningeal vessel (present on the right side only); in accordance with Smith's (1928) metrical analysis of the Kabwe endocast; Holloway's (1980) study of the Ngandong endocast; and Holloway et al. (2004) for their recent study of hominin brain endocasts. The measurement of greatest breadth was taken at a point immediately inferior to a circular deformation and posterior and superior to the posterior branch of the middle meningeal vessel in the posterior temporoparietal region. This area is not present on the left hemisphere, so its contour was estimated on the adjacent areas and the homologous area on the right hemisphere. Maximum length was obtained using

Weil's (1929) and Holloway et al's (2004) refined approaches by taking the maximum chord between the frontal and occipital lobes on each side and drawing a perpendicular line from each endpoint to the median plane.

Figure 3

View of the right hemisphere showing detail of anterior (a), middle (m), and posterior (p) branches of middle meningeal vessels. The circular growth (x) lying on the parietal lobe is a meningioma. Dashed lines: sutural markings. Stippling: reconstructed areas

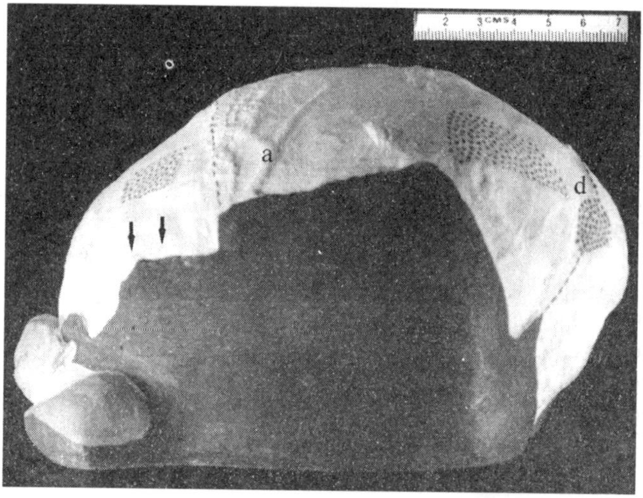

Results

Deformation within the parietal region

The circular deformation lying in the parietal region of the right hemisphere (Figure 3) represents a pathological condition. This trace of what appears to have been an abnormal growth that eroded a circular area of the inner table of the skull has been partially diagnosed as a probable meningioma (personal communication, E.E. Duda, MD). Meningiomas are tumours that arise in arachnoid tissues embedded in the outer layer of the dura (Jónsdóttir et al., 2003: 232). Most commonly they occur in the cranium near the superior sagittal sinus and are usually slow growing and do not metastasize to other organs and tissues (op. cit.). Meningioma constitute between 12–19 per cent of all intracranial tumours (Traub, 1961; quoted in Jónsdóttir et al., 2003). Destructive meningiomas are the least common.

Morphological observations

Vasculature: A stout ridge representing the anterior branch of the middle meningeal vessels lies posterior to the coronal suture and extends towards vertex, ending 45 mm (right) and 36 mm (left) from midline (Figures 3–5). Short side branches extend anteriorly and posteriorly from the vessel on both hemispheres. On the right, the posterior branch of the middle meningeal vessel arises from the broken inferior border of the squamous temporal and extends towards a reconstructed area of the parietal. Along its course, the middle branch arises and is shortly obliterated by the circular deformation of the parietal surface. Superior to the deformation, the middle meningeal branch continues

Figure 4

View of the left hemisphere showing detail of the anterior branch of the middle meningeal vessels (a), and pars triangularis and opercularis (arrows), ascending ramus of the Sylvian fissure on the inferior frontal gyrus (other abbreviations as in Figures 3 and 5), and swell depression (d)

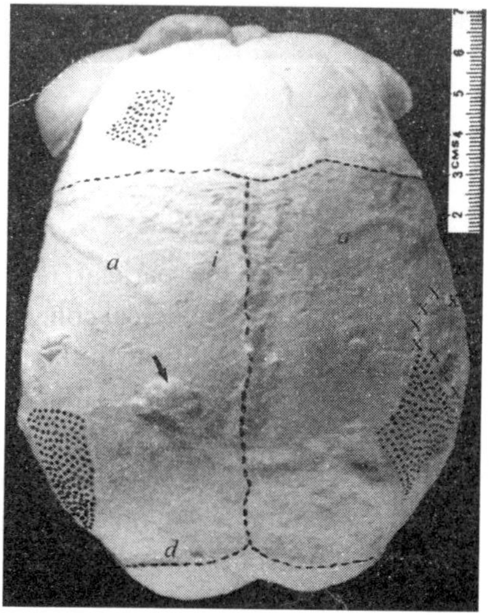

medially for a short distance, appearing to branch approximately 30 mm from the sagittal suture. On the left, artefacts, deficiencies, and reconstruction leave no trace of posterior or middle branches of the middle meningeal vessels.

The left and right transverse (lateral) sinuses lie in their usual position between posterior aspects of occipital lobe and cerebellum. In a.m. *Homo sapiens*, the superior longitudinal sinus commonly ends in the right transverse sinus (Smith, 1907: 575) which appears to be the situation in the Saldanha endocast (see Table 3 and Figure 6).

Petalia: In most modern humans, the left occipital pole extends further posteriorly to form a deeper impression (petalia[1]) on the left than on the right side of the occipital bone. Similarly, in most modern humans, the right frontal lobe is wider and extends further anteriorly (see Table 2). As in a.m. *Homo sapiens*, a counter-clockwise "torque" (LeMay, 1977: 251) characterizes the form of the Saldanha endocast (Figure 5).

Figure 5

View of superior aspect of the Saldanha endocast showing right and left middle meningeal vessels, arachnoid granulations (arrow), pre-(sub-) lambdoid depression (d), inferior parietal lobule (i), areas of reconstruction, and artefactual erosions (other abbreviations as in Figure 2)

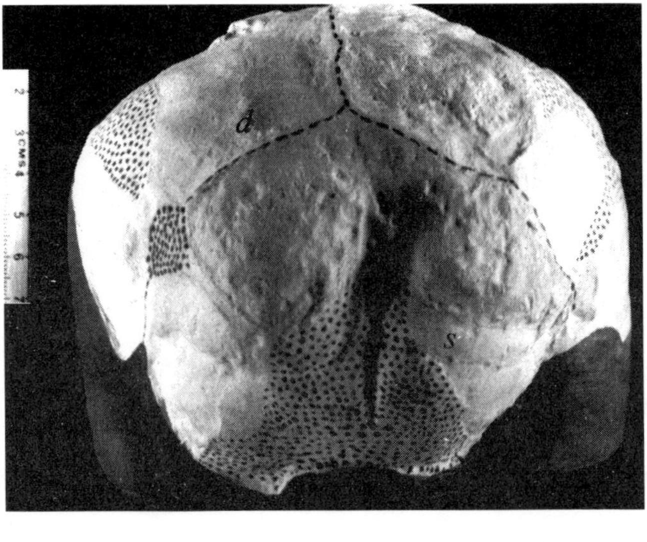

Table 3

Frequencies of hemispheric asymmetries in petalia and direction of major branch of lateral sinus in modern and fossil hominins

Taxon	*Lateral sinus*			*Petalia (%)*					*References*
	M.B. (%)				*Frontal*		*Occipital*		
	n	*lt*	*rt*	*n*	*lt*	*rt*	*lt*	*rt*	
a.m. *Homo sapiens*	123	15	50	130*	14	66	77	12.5	Hochberg and LeMay, 1975: 220; LeMay, 1977: 248
				14	0	100	77	23	Holloway and DeLacoste, 1982: 105
Homo neanderthalensis	NA			3#	0	33	66	0	LeMay, 1976: 361
Homo heidelbergensis	2&	0	100	2	0	50	50	0	Smith, 1928: 35; this study
Homo (sp. indet.) (Solo)	5	20	80	5	20	80	80	0	Holloway, 1980: 292
Homo erectus (Indonesia)	NA			2	0	0	50	50	LeMay, 1976: 361
All Homo erectus groups (including Solo)	NA			occ19 fro13	8	92	89	11	Holloway and DeLacoste, 1982: 105

(NA = not available; sexes pooled; M.B. = major branch of transverse sinus; * = right-handers only, the left-handed sample being virtually equally distributed between left, right, and no major direction of M.B.; # = specimens in *Homo neanderthalensis* sample: Neanderthal, La Chapelle, La Quina; & = specimens in *Homo heidelbergensis* sample: Kabwe, Saldanha).

Convolutions: In human skulls, the temporal and orbital regions, i.e., those lacking on the Saldanha endocast, are regions that retain the greatest number of fissural markings (Connolly, 1950: 323; Hirschler, 1942: 124). A few *gyri* (convolutions) and fissures, however, show on the frontal and parietal regions of the Saldanha calotte. Although positive identification of these scattered features may be moot, a brief description of tentatively identified features may be of interest for

Figure 6

View of the posterior aspect of the Saldanha endocast showing parts of the transverse sinuses (s) (clearly visible of the right side only), occipital poles and small part of the cerebellum, and lateral asymmetry in the conformation of parietooccipital region, probably an artefact of the reconstruction (abbreviations as in Figure 3)

comparison with other endocranial material (and for inferences concerning behavioural capacities in transitional hominins).

Arachnoid granulations spread widely at bregma and along the midsagittal line (Figure 4, arrow), obscuring any possible cerebral digitations or crests on the superior parietal surface. On the left, 10 mm anterior to the lambdoid suture, two smooth elevations appear to represent the inferior parietal lobule. The lambdoid suture appears to be raised relative to a depression that lies immediately anterior to it.

Impressions of occipital, cerebellar, and right posterior temporal *gyri* have been obliterated by erosion (Figure 6). The right frontal region (Figure 7, arrow) shows two elevations on its anterior surface in the position of the middle frontal *gyrus* (*frontalis medialis* 1 and 2 of Hirschler, 1942). On the left hemisphere, a smooth elevation bordered laterally by a reconstructed surface lies in the superior frontal *gyrus* (*frontalis superior* 3). More inferiorly, a slight elevation might represent the left middle frontal convolution, but the eroded quality of this surface makes such an interpretation uncertain. Medial to the broken lateral border of the left frontal lobe, two transverse elevations (Figure 7, arrows) might represent the upper parts of *pars triangularis* and *pars opercularis* of the inferior frontal *gyrus* and the *sulcus* lying between these eminences, the ascending ramus of the lateral cerebral sulcus (*Sylvian* fissure).

Figure 7

Anterior view of the Saldanha endocast showing gyri frontalis medialis 1 and 2 (1, 2) on the right hemisphere and gyri frontalis superialis 3 (3) on the left. Shaded areas ventral to the hemisphere represent frontal sinuses

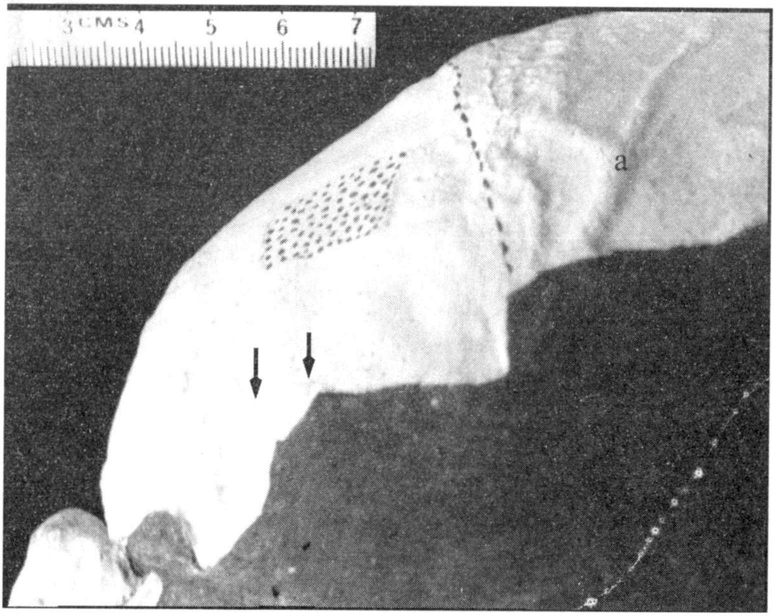

General form and volume: Maximum length for each cerebral hemisphere of the Saldanha endocast measures 157 mm from frontal pole to occipital pole (differing from Drennan's, 1953b: 10, report on an earlier endocast of the original Saldanha calotte, measuring 165 mm in length). According to an average of six measurements, the equality of length in the two hemispheres is not reflected by an appearance of symmetry. As the right frontal lobe extends anteriorly and the left occipital posteriorly (Figure 5), the hemispheres are displaced with respect to one another by a length of 2 mm (see "*Petalia*," above). As in the Kabwe and Ngandong endocrania, maximum breadth occurs at the posterior temporoparietal region (Figure 6). In the Saldanha endocranium, maximum breadth measures 124 mm. The endocranial length-breadth index of the Saldanha endocast is 0.790.

We agree with Drennan (1953b: 10) that in vertical profile on the Saldanha endocast, the cerebellum lies well back and is barely overlapped by the occipital poles above it (Figures 3 and 4). We believe that a portion of the right side of *planum nuchale* has been misaligned in reconstruction of the Saldanha skull, in that it has been tilted too much horizontally. It is specifically that portion which produces an appearance of the occipital pole being pressed upwards by a more inferiorly placed cerebellum, leading Drennan to the conclusion that the occipital pole is "rather feebly developed" (Drennan, 1955: 630). Singer was involved in this reconstruction and disagreed with Drennan on the positioning of the *planum nuchale* on the right side. We believe that a more accurate alignment would remove the "inferior status" of the region and place it more in line with that of modern humans.

An endocranial capacity of "greater than 1,200 cm^3" was estimated on the basis of water displacement method (Drennan, 1953a: 793; 1953b: 10; Holloway et al., 2004). The derivation of a more specific value here seems neither feasible nor necessary, in light of the wide range of variation in brain size of Middle and Upper Pleistocene hominins of *erectus*, Neanderthal, and archaic modern grades (Table 2).

Morphological comparisons

The pattern of middle meningeal vasculature on the Saldanha endocast resembles that of Guiffridi-Rugieri's (1913: 404) class IIb, in which a stout anterior (frontoparietal or bregmatic) branch gives rise to a posterior (lambdoidal) branch, which in turn gives rise to a middle (obelionic) branch. In contrast, the Kabwe endocast bears a class I pattern in which the anterior branch gives rise to both the posterior and middle branches.

Both patterns (among others) are characteristic of a.m. *Homo sapiens* (see Hirschler, 1942: 81), the salient feature being the relative robusticity of the anterior branch with respect to the temporal and posterior branches. In endocrania of (the former) *Sinanthropus* and *Pithecanthropus* described by Weidenreich (1938: 13), the anterior branch is rather diminutive compared to the temporal (including middle and posterior) branches that comprise the predominant middle meningeal vessels. In Neanderthal endocrania, middle meningeal patterns appear generally to resemble the more “advanced” pattern of stout frontoparietal branch, while the classic Neanderthal specimen shows stouter posterior branch and diminutive anterior branch (op. cit.: 10). The magnitude of variability, as well as the general trend toward increasing robusticity of the anterior branch of the middle meningeal vessels, is apparent in Saban’s (1983) illustrations of middle meningeal grooves in fossil hominins.

The Saldanha endocast resembles the Kabwe endocast with regard to the pre-lambdoid depression (Smith, 1928: 54). Smith has attributed this feature to the lunate sulcus, a diagnosis not accepted here. As Hirschler (1942: 2) has pointed out, this furrow is caused by a thickening of the neighbouring occipital and parietal bones along their borders and cannot be regarded as corresponding to the lunate sulcus. Since the Saldanha and Kabwe cranial capacities do not differ from the low end of the modern human range (see Table 2), there is no reason to expect the presence of such a conspicuous or anteriorly placed lunate sulcus in either of these “transitional” brains. In a.m. humans, the lunate sulcus, if present, lies much more inferiorly near the occipital pole (Warwick and Williams, 1973: 926). The phylogenetic significance of the pre-lambdoid furrow (Figure 5) is unclear. A “sub”-lambdoid depression was observed in a modern Chinese endocast (Hirschler, 1942: 79), and the frequency of this endocranial trait is not reported for a.m. populations of *Homo sapiens*.

In lateral profile of Kabwe, the frontal region appears to be rounded; the middle fossa shallow and the posterior fossa extremely shallow, long, and low in the Kabwe endocranium (Drennan, 1955: 627, 630; Pycraft, 1928: 11). Superimposed X-rays of lateral profiles of Saldanha and Kabwe (Figure 2) suggest that, although the anterior and middle fossae are too obscure to diagnose, the posterior fossae in the two specimens differ. The Saldanha specimen shows a somewhat less shallow posterior fossa than does Kabwe, due to a less horizontal orientation of the *planum nuchale*. Even with the possible misalignment in the reconstruction of the posterior region on the right (see pp. 15–16), the X-ray indicates that this region of the Saldanha endocranium is somewhat more rounded than that of Kabwe.

While length and breadth in the Saldanha endocranium measure 157 mm and 124 mm, respectively, those measurements in Kabwe are 170 and 135. Although the Saldanha endocast represents a brain somewhat smaller in overall dimensions (even prior to ca. 4% shrinkage), the endocranial length-breadth indices of Saldanha and Kabwe (0.790 and 0.794, respectively) differ by only 0.5 per cent. A trend towards increasingly rounded brains (signified by higher length-breadth indices also noted by Holloway et al., 2004) appears to characterize the phylogenetic series from earlier (*erectus*) through indeterminate (Solo) to later (Neanderthal) grades. Interestingly, a modern endocranial index value of 0.787, characterizing a modern (Chinese) population lies between Neanderthal and Solo values of 0.793 and 0.774, respectively. Three explanations for this (not mutually exclusive) come to mind:

1. Although endocranial characters are reported in scattered publications, ranges of variation of craniometric values for various hominin populations (including modern) are not extensively reported. It is possible that the size and shape peculiarities of the known fossil hominin endocrania represent a biased sample showing an artefactual trend towards brachycephalization (Weidenreich, 1945: 49) and, concomitantly, "brachyencephalization." (In general, ectocranial indices are slightly lower than endocranial indices of the same skulls and brains are concomitantly somewhat more rounded than the

external form of the skulls that contain them; e.g., Weidenreich, 1945: 16. See "Petralona" index value, Table 2.) Of course, not all evidence supports the view that a major trend during hominin evolution has been towards brachycephaly (see, e.g., Huizinga and Slob, 1965: 300; Kappers, 1935: Tables).

2. The "trend" in hominin phylogeny towards increased brachyencephaly might be inoperative in a.m. *Homo sapiens*. A relaxation of selection pressure for brachyencephaly (or some pleiotropically linked character) in modern human populations may have produced a wider range of variation in brain shape than was present in fossil hominins, so that a greater range of brain shape exists instead of a "trend."
3. A general trend towards brachyencephaly might have occurred concomitantly with a possible change in maturational processes (Holloway et al., 2004; Trinkaus, 1984: 273; Zihlman, 1982: 953), so that a genuine difference in brain shape might be reflected in the trend towards more rounded brains in the hominin fossil record. Weidenreich (1945: 415) has discussed the direct relationship between relative brain size and concomitant sphericity, positing that fetal growth rates and not fetal features *per se*, are retained in evolving hominins. The effect of fetalization of hominin growth rates may be manifest in brachyencephaly in a.m. *Homo sapiens*, but this is not clear from the available evidence.

With regard to cerebral asymmetries, the Saldanha and Kabwe specimens show both similarity and difference. While the Saldanha cerebrum shows the typical pattern of counter-clockwise torque of frontal and occipital petalia, asymmetry in petalia is absent in Kabwe. The two specimens show a dextral bending of the superior longitudinal sinus into the transverse sinus. That there exists a difference in lateral asymmetry in this sample of two fossil hominin representatives of *Homo sapiens rhodesiensis* is not surprising, since lack of asymmetry and, even, reversal of the typical pattern characterizes a substantial minority of modern humans (Table 3).

Sulcal patterns of the Saldanha endocast are not diagnostic, outside of the tentative identification of Broca's area on the left inferior frontal *gyrus* and a well-developed left inferior parietal lobule. LeMay

(1975: 12) has identified a longer left Sylvian fissure (also associated with the development of the left hemisphere's specialization for language – see Conroy, 2002 for more discussion) in the Neanderthal from La Chappelle-aux-Saint.

Discussion

Judging from the fossil evidence, late Middle to early Upper Pleistocene hominins in sub-Saharan Africa did not differ significantly in major craniometric features from that of a.m. *Homo sapiens*. Similarly, features of brain size and gross conformation consistently fall within modern normal limits to the extent that available evidence permits comparison (see Holloway et al., 2004 on endocast comparisons). In terms of brain size, length-breadth dimensions, and endocranial length-breadth indices, the Saldanha endocast, like the Kabwe endocast, lies within the rather wide range that characterizes a.m. *Homo sapiens*. In conjunction with associated archaeological and contextual evidence, a few inferences concerning the behaviour of these "transitional" sub-Saharan hominins can be tendered.

Neither brain size nor endocranial morphology differ from the conditions that exist in the a.m. human range, so available evidence does not challenge the suggestion that technological and linguistic "capacity" for example, in the late Middle and early Upper Pleistocene southern African hominins followed a pattern towards that of a.m. human populations. As it is generally understood, the presence of particular morphological cerebral features does not necessarily indicate specific behavioural functions (Holloway et al., 2004); the following inferences concerning laterally specialized capacity are provisional. Especially in light of evidence that cultural development was at an early stage during the late Middle and early Upper Pleistocene, rudimentary performances, at most, might be inferred in the context of functionally lateralized capacities for language and right-handedness.

Right-handedness is suggested by the counter-clockwise torque (LeMay, 1977: 246) of the Saldanha endocast. Technological capacity requires manual specialization that relies on hemispheric specialization. In modern humans, the incidence of right-handedness is probably in the range of 90–92 per cent (Hardyck and Petrinovich, 1977: 391). Interestingly, an approximate frequency of 90 per cent right-handedness appears to have remained more or less constant since "pre 3000 B.C." and probably since as long ago as 15,000 B.C. (Coren and Porac, 1977: 631). Of course, this is not nearly so long ago as the transitional populations considered here, but the Acheulian tools tentatively associated with both the Saldanha cranial fragments and numerous elephant bones support the interpretation that dexterity, if not dextrality, was a non-random occurrence in these transitional hominins. Absence of evidence for *petalia* asymmetry in Kabwe does not challenge this view, since the frequency of hemispheric morphological asymmetry is not as high as that of behavioural lateral specialization even in a.m. *Homo sapiens*.

Although the relationship between lateralized hemispheric functions and asymmetries in the hemispheres' gross morphology has not been conclusively demonstrated (Witelson, 1983: 130), such a relationship between form and function is widely regarded as necessary, albeit complex. Although the extent of the social organization, as well as technological capacity, of the Saldanha individual remains unknown, a pronounced development of the left inferior frontal *gyrus* in the region known as Broca's area suggests development of some capacity for speech production. Specifically, this region of the inferior frontal *gyrus* of the left hemisphere is considered to comprise Broca's area, a region associated with the motor element of speech (Holloway et al., 2004; Warwick and Williams, 1973: 924). The accompanying evidence of a development of the inferior parietal lobule suggests that the Saldanha brain was also capable of comprehending language (Tobias, 1981: 40), since the inferior parietal lobule is held to be an "association area of association areas" (Davenport, 1977: 81; Falk et al., 2000), and cross-modal association is held to be a prerequisite for linguistic capacity.

The existence of developed cultural capacities by transitional *Homo sapiens* cannot accurately be explained. However, as suggested by Holloway et al. (2004), a mosaic pattern of the evolution of the brain could have facilitated such events. Selective pressures on various parts of the brain associated with specific functions coupled with neurological patterning and ontogenetic development over time may have paved a way for cultural capacities. Selective forces acting on upon different regions of the brain may have influenced linguistic, technological, and social behaviour in the earliest sapient hominins.

Conclusions

Absolute brain size may have increased in hominins over geological time, but determination of relative brain size involves estimation of body size and sexual dimorphism, as well as possible developmental variability. Moreover, the variance present in brain sizes measured over geological time on the basis of the fossil record is high enough to shed doubt on a simple view that brain size increased gradually as hominin populations evolved (Rightmire, 1981: 245). There is nothing about the Saldanha endocast to rule out an interpretation of its brain as that of a sapient hominin, comparable in morphometric terms, at least, with members of a.m. *Homo sapiens*.

Similarities between the Saldanha endocast and the more complete Kabwe specimen in vascular pattern, size, shape, and asymmetry suggest that both brains, representing the brain of *Homo heidelbergensis*, lie within the range of a.m. human values. The similarity between the endocasts of Kabwe and Saldanha is striking, but, as with the skeletal material, notable differences exist. The differences include details of middle meningeal pattern, *petalia* conformation, and overall size.

The differences no doubt reflect the range of variation, and perhaps sexual dimorphism, in a broadly contemporaneous late Middle to early Upper Pleistocene hominin population in sub-Saharan Africa. This population is characterized by brain sizes within the range of 1,000–1,325 cm^3. Brain asymmetries that are associated or, at least, not opposed to the left-right pattern of hemispheric asymmetry typical of a.m. *Homo sapiens*, may suggest behavioural capacity that is no doubt culturally patterned. The extent of social, communicative, and technological elaboration, of course, remains an intriguing mystery in this transitional population of sub-Saharan African *Homo sapiens*.

Acknowledgement

Without the efforts of the late Dr Ronald Singer, this review article would not have been possible; we therefore dedicate this article to his life. We thank Drs Russell Tuttle, David Tracer, and Ms Julie Lesnik for reviewing the earlier version of this manuscript and for providing constructive criticism and suggestions. We also thank Ms Kate Ulmer and Connie Turner for the digitization of the photographs and illustrations. We thank the editors of this volume and the three anonymous reviewers for providing insightful comments from which the quality of this manuscript benefited tremendously.

Note

1. "*Petalia*" refers to a protrusion of one hemisphere compared to the other, reflected on the internal aspect of the squamous portion of the occipital, parietal, or frontal bones. Occipital *petalia*, then, refers to the asymmetry of the two superior fossae on the internal aspect of the squamous part of the occipital bone produced by asymmetry of the occipital lobes of the cerebral hemispheres.

References

Adefris, T. 1992. A description of the Bodo cranium: an archaic *Homo sapiens* cranium from Ethiopia. Ph.D. Dissertation, New York University, New York.

Asfaw, B., Gilbert, W.H., Beyene, Y., Hart, W.K., Renne, P.R., WoldeGabriel, G., Vrba, E.S., and White, T.D. 2002. Remains of *Homo erectus* from Bouri, Middle Awash, Ethiopia. *Nature* 416(21): 317–319.

Bräuer, G. 1984. A craniological approach to the origin of anatomically modern *Homo sapiens* in Africa and implications for the appearance of modern Europeans. In F.H. Smith and F. Spencer (eds) *The Origins of Modern Humans: A World Survey of the Fossil Evidence*. pp. 327–410. New York: Alan R. Liss.

Bräuer, G. 1992. Africa's place in the evolution of *Homo sapiens*. In G. Bräuer and F.H. Smith (eds) *Continuity or Replacement? Controversies in Homo sapiens Evolution*. pp. 83–98. Rotterdam: Balkema.

Bräuer, G., Groden, C., Delling, G., Kupczik, K., Mbua, E., and Schultz, M. 2003. Pathological Alterations in the Archaic Homo sapiens Cranium From Eliye Springs, Kenya. *Am. J. Phys. Anthropol.* 120(2): 200–204.

Bräuer, G. and Stringer, C.B. 1997. Models, polarization, and perspectives on modern human origins. In G.A. Clark and C. Willermet (eds) *Conceptual Issues in Modern Human Origins Research.* pp. 191–201. New York: Aldine de Gruyter.

Butzer, K.W. 1973. Reevaluation of the geology of the Elandsfontein (Hopefield) Site, South-west Cape, South Africa. *S. Afr. J. Sci.* 69: 234–238.

Butzer, K.W., Beaumont, P.B., and Vogel, J.C. 1978. Lithostratigraphy of Border Cave, KwaZulu, South Africa: a Middle Stone Age sequence beginning c. 195,000 B.P. *J. Arch. Sci.* 5: 317–341.

Clark, J.D., Oakley, K.P., Wells, L.H., and McClelland, J.A.C. 1950. New studies on Rhodesian man. *J. Roy. Anthrop. Inst.* 77: 7–32.

Clark, J.D. et al. 2003. Stratigraphic, chronological, and behavioral contexts of Pleistocene *Homo sapiens* from Herto, Middle Awash, Ethiopia. *Nature* 423: 747–752.

Clark, R.J. 1976. New cranium of *Homo erectus* from Lake Ndutu, Tanzania. *Nature* 262: 485–487.

Clark., R.J. 1990. The Ndutu cranium and the origin of *Homo sapiens. J. Hum. Evol.* 19: 699–736.

Compton, J.S. and Franceschini, G. 2005. Holocene geoarchaeology of the Sixteen Mile Beach barrier dunes in the Western Cape, South Africa. *Quatern. Res.* 63: 99–107.

Connolly, J.C. 1950. *External Morphology of the Primate Brain.* Springfield, IL: Charles C. Thomas.

Conroy, G.C. 2002. Speciocity in the early *Homo* lineage: too many, too few, or just about right? *J. Hum. Evol.* 43: 759–766.

Conroy, G.C., Jolly, C.J., Cramer, D., and Kalb, J.E. 1978. Newly discovered fossil hominid skull from the Afar depression, Ethiopia. *Nature* 276: 67–70.

Coren, S. and Porac, C. 1977. Fifty centuries of right-handedness: the historical record. *Science* 198: 631–632.

Davenport, R.K. 1977. Cross – modal perception: a basis of language? In D.M. Rumbaugh (ed.) *Language Learning by a Chimpanzee: The LANA Project.* pp. 73–83. New York: Academic Press.

Day, M.H., Leakey, M.D., and Magori, C. 1980. A new hominid fossil skull (L.H. 18) from the Ngaloba Beds, Laetoli, northern Tanzania. *Nature* 284: 55–56.

Drennan, M.R. 1953a. The Saldanha skull and its associations. *Nature* 172: 791–793.

Drennan, M.R. 1953b. A preliminary note on the Saldanha skull. *S. Afr J. Sci.* 50: 7–11.

Drennan, M.R. 1955. The special features and status of the Saldanha skull. *Am. J. Phys. Anthropol.* 13: 625–634.

Dreyer, T.F. 1936. The endocranial cast of the Florisbad skull – a correction. *Sool. Navors. Nas. Mus.* (Bloemfontein) 1(5): 21 – 23.

Dreyer, T.F. 1938. The fissuration of the frontal endocranial cast of the Florisbad skull compared with that of the Rhodesian skull. *Zts. f. Rassenkunde* 8: 193–198.

Eldredge, N. and Tattersall, I. 1982. *The Myths of Human Evolution.* New York: Columbia University Press.

Falk. D., Redmond, J.C., Conroy, G.C., Recheis, W., Weber, G.W., and Seidler, H. 2000. Early hominid brain evolution: a new look at old endocasts. *J. Hum. Evol.* 38: 695–717.

Gabunia, L., Vekua, A., and Lordkipanidze, D. 2001. New human fossils from Dmanisi, eastern Gorgia. *Archaeol. Ethnol. Anthropol. Eurasia* 2(6): 128–139.

Guiffridi-Ruggeri, V. 1913. Uber die endocranischen Furchen der Arteria meningea media beim Menschen. *Zeitschr. f. Morph. Anthrop.* 15: 401–412.

Grün, R. and Beaumont, P. 2001. Border Cave revisited: a revised ESR chronology. *J. Hum. Evol.* 40: 467–482.

Grün, R., Brinks, J.S., Spooner, N.A., Taylor, L., Stringer, C.B., Franciscus, R.G., and Murray, A.S. 1996. Direct dating of Florisbad hominid. *Nature* 382: 500–501.

Haile-Selassie, Y., Asfaw, B., and White, T.D. 2004. Hominid cranial remains from Upper Pleistocene deposits at Aduma, Middle Awash, Ethiopia. *Am. J. Phys. Anthropol.* 123(1): 1–10.

Hardyck, C. and Petrinovich, L.F. 1977. Left-handedness. *Psychol. Bull.* 84: 385–404.

Hendey, Q.B. 1969. Quaternary vertebrate fossil sites in the southern Cape Province. *S. Afr. Archeol. Bull.* 24: 96–105.

Hirschler, P. 1942. *Anthropoid and Human Endocranial Casts.* Amsterdam: N.V. Noord – Hollandsche Uitgevers Maatschapij.

Hochberg, F.H. and LeMay, M. 1975. Arteriographic correlates of handedness. *Neurology* 25: 218–222.

Hofman, M.A. 1983. Encephalization in hominids: evidence for the model of punctuationalism. *Brain. Behav. Evol.* 22: 102–117.

Holloway, R.L. and de La Coste-Laremonde, M.C. 1982. Brain endocast Symmetry in Pongids and Hominids: Some preliminary findings in the paleontology of cerebral dominance. *Am. J. Phys. Anthropol.* 60: 505–5165.

Holloway, R.L. and de La Coste-Laremonde, M.C. 1986. Sexual dimorpohisans in the human corpus collosum: an extensive and replication study. *Human Neurobiology*. 5: 85–91.

Holloway, R.L. 1980. Indonesian "Solo" (Ngandong) endocranial reconstructions: some preliminary observations and comparisons with *Neanderthal* and *Homo erectus* groups. *Am. J. Phys. Anthropol.* 53: 285–295.

Holloway, R.L. 1982. *Homo erectus* brain endocasts: volumetric and morphological observations with some comments on cerebral asymmetries. 1[er] *Congres Internationale Paleontologie Humaine*, Nice. J. Piveteau, H. deLumley, and M. deLumley (eds) pp. 355.

Holloway, R.L., Broadfield, D.C., and Yuan, M.S. 2004. T*he Human Fossil Record, Volume Three: Brain Endocasts - The Paleonurological Evidence*. New York: Wiley-Liss.

Howell, F.C. 1978. Hominidae. In V.J. Maglio and B.S. Cooke (eds) *Evolution of African Mammals*. pp. 154–248. Cambridge, MA: Harvard University Press.

Huizinga, J. and Slob, A. 1965. Progressive brachycephalization reproduction and head form in the Netherlands. A preliminary report. *Koninkl. Nederl. Akademie von Wetenschappen – Amsterdam*. Proc. Ser. C. 68: 4. pp. 298–301.

Jónsdóttir, B., Ortner, D.J., and Frohlich, B. 2003. Probable destructive meningioma in an archaeological adult male skull from Alaska. *Am. J. Phys. Anthropol.* 122: 232–239.

Kappers, C.U.A. 1935. The degree of the changes in the cephalic index correlated with age and environment. *Koninkligkr. Akademie Van Wetenschappen Te Amsterdam* 38(9): 3–15.

Klein, R.G. 1973. Geological antiquity of Rhodesian man. *Nature* 244: 311–312.

Klein, R.G. 1978. The fauna and overall interpretation of the "Cutting 10" Acheulean site at Elandsfontein (Hopefield), Southwestern Cape Province, South Africa. *Quatern. Res.* 10: 69–83.

Klein, R.G. 1983. The Stone Age prehistory of southern Africa. *Ann. Rev. Anthropol.* 12: 25–48.

Klein, R.G., Avery, G., Cruz-Uribe, K., and Steele, T.E. 2007. The mammalian fauna associated with an archaic hominin skullcap and later Acheulean artifacts at Elandsfontein, Western Cape Province, South Africa. *J. Hum. Evol.* 52: 164–186.

Klein, R.G. and Cruz-Uribe, K. 1991. The bovids from Elandsfontein, South Africa, and their implications for the age, paleoenvironment, and origins of the site. *Afr. Archaeol. Rev.* 9: 21–79.

Leakey, M.D. and Hay, R.L. 1982. The chronology of the fossil hominids of Tanzania. *1[er] Congres International Paleontologie Humaine*, Nice. J. Piveteau, H. deLumley, and M. deLumley (eds). pp. 753–765.

LeMay, M. 1975. The language capability of Neanderthal man. *Amer. J. Phys. Anthropol.* 42: 9–14.

LeMay, M. 1977. Asymmetries of the skull and handedness. Phrenology revisited. *J. Neurol. Sci.* 32: 243–253.

Magori, C.C. and Day, M.H. 1983. Laetoli Hominid 18: an early *Homo sapiens* skull. *J. Hum. Evol.* 12: 747–753.

Manega, P.C. 1993. Geochronology, geochemistry and isotopic study of the Plio-Pleistocene hominid sites and the Ngorongoro volcanic highlands in northern Tanzania. Ph.D. Dissertation, University of Colorado, Boulder, CO.

Martinez, I. and Arsuaga, J.L. 1997. The temporal bones from Sima de los Huesos Middle Pleistocene site (Sierra de Atapuerca, Spain). A phylogenetic approach. *J. Hum. Evol.* 33: 283–318.

Mehlman, M.J. 1984. Archaic *Homo sapiens* at Lake Eyasi, Tanzania: recent misrepresentations. *J. Hum. Evol.* 13: 487–501.

Morant, G.M. 1928. Studies of Paleolithic man. III. The Rhodesian skull and its relations of Neanderthaloid and modern types. *Ann. Eugen.* 3: 337–360.

Mturi, A.A. 1976. New hominid from Lake Ndutu, Tanzania. *Nature* 262: 484–485.

Murrill, R.I. 1968. *Cranial and Post-cranial Skeletal Remains from Easter Island.* Minneapolis: University of Minnesota Press.

Olivier, G. and Tissier, H. 1975. Determination of cranial capacity in fossil men. *Am. J. Phys. Anthropol.* 43: 353–362.

Partridge, T.C. 1982. The chronological positions of the fossil hominids of southern Africa. *1[er] Congres Internationale Paleontologie Humaine,* Nice. J. Piveteau, H. deLumley and M. deLumley (eds). pp. 617–675.

Pycraft, W.P. 1928. Rhodesian man. Description of the skull and other human remains from Broken Hill. In W.P. Pycraft (ed.) *Rhodesian Man and Associated Remains*. pp. 1–51. London: British Museum of Natural History.

Rightmire, G.P. 1979. Remains of *Homo erectus* from Beds II and IV, Olduvai Gorge, Tanzania. *Am. J. Phys. Anthropol.* 51: 99–116.

Rightmire, G.P. 1981. Patterns in the evolution of *Homo erectus. Paleobiology* 7: 241–246.

Rightmire, G.P. 1993. *The Evolution of Homo erectus: Comparative Anatomical Studies of an Extinct Human Species*. Cambridge: Cambridge University Press, p. 260.

Rightmire, G.P. 1998. Human evolution in the Middle Pleistocene: the role of *Homo heidelbergensis. Evol. Anthropol.* 6: 218–227.

Rightmire, G.P. and Deacon, H.J. 2001. New human teeth from Middle Stone Age deposits at Klasies River, South Africa. *J. Hum. Evol.* 41: 535–544.

Saban, R. 1983. Anatomie et Evolution des Veines Meningees Chez les Hommes Fossiles. *Memoires de la Section des Sciences*. Paris: E.M.S.B.-C.T.H.S.

Seidler, H., Falk, D., Stringer, C., Wilfing, H., Müller, G.B., zur Nedden, D., Weber, G.W., Reicheis, W., and Arsuaga, J-L. 1997. A comparative study of stereolithographically modeled skulls of Petralona and Broken Hill: implications for future studies of middle Pleistocene hominid evolution. *J. Hum. Evol.* 33: 691–703.

Sillen, A. and Morris, A. 1996. Diagenesis of bone from Border Cave: implications for the age of the Border Cave hominids. *J. Hum. Evol.* 31: 499–506.

Singer, P. and Wymer, J. 1982. *The Middle Stone Age at Klasies River Mouth in South Africa*. Chicago: University of Chicago Press.

Singer, R. 1954. The Saldanha skull from Hopefield, South Africa. *Am. J. Phys. Anthropol.* N.S. 12: 345–362.

Singer, R. and Crawford, J.R. 1957. The significance of the archaeological discoveries at Hopefield, South Africa. *J. Roy. Anthrop. Inst.* 88: 11–19.

Singer, R. and Wymer, J. 1968. Archaeological investigations at the Saldanha skull site in South Africa. *S. Afr. Archaeol. Bull.* 23: 63–74.

Smith, C.E. 1907. On the asymmetry of the caudal poles of the cerebral hemispheres and its influence on the occipital bone. *Anat. Anz.* 30: 574–578.

Smith, G.E. 1928. Endocranial cast obtained from the Rhodesian skull. In W.P. Pycraft (ed.) *Rhodesian Man and Associated Remains*. pp. 52–63. London: British Museum of Natural History

Stringer, C.B. 1992. Reconstructing recent human evolution. *Philos. Trans. R. Soc. Lond. (Biol.)* 337: 217–224.

Stynder, D.D., Moggi-Cecchi, J., Berger, L.R., and Parkington, J.E. 2001. Human mandibular incisors from the late Middle Pleistocene locality of Hoedjiespunt 1, South Africa. *J. Hum. Evol.* 41: 369–383.

Thackeray, J.F. 2002. Palaeoenvironmental change and re-assessment of the age at Late Pleistocene deposits at Die Kelders cave, South Africa. *J. Hum. Evol.* 43: 749–753.

Tobias, P.V. 1981. Evolution of human brain, intellect, and spirit. 1st Abbie Memorial Lecture, 1979. University of Adelaide, South Australia.

Trinkaus, E. 1984. Western Asia. In F.H. Smith and F. Spencer (eds) *The Origins of Modern Humans: A World Survey of the Fossil Evidence*. pp. 251–293. New York: Alan R. Liss.

Vrba, E.S. 1982. Biostratigraphy and chronology, based particularly on bovidae, of southern hominid-associated assemblages: Makapansgat, Sterkfontein, Taung, Kromdraai, Swartkrans also Elandsfontein (Saldanha) Broken Hill (now Kabwe), and Cave of Hearths. *1er Congres Internationale de Paleontologie Humaine*, Nice. J. Piveteau, H. deLumley, and M. deLumley (eds). pp. 707–752.

Walker, A. and Leakey, R.E.F. 1978. The hominids of East Turkana. *Sci. Am.* 239: 54–66.

Warwick, R. and Williams, P.L. 1973. *Gray's Anatomy*, 35th British edition. New York: W.B. Saunders.

Weidenreich, F. 1938. The ramification of the middle meningeal artery in fossil hominids and its bearing upon phylogenetic problems. *Palaeontol. Sini.* N.S. D, 3: 1–16.

Weidenreich, F. 1941. The brain and its role in the phylogenetic transformation of the human skull. *Trans., Am. Philos. Soc.* N.S. 31: 321–442.

Weidenreich, F. 1943. The skull of *Sinanthropus pekinensis*: a comparative study of a primitive hominid skull. *Palaeontol. Sin.* N.S. D, 10, W.S. 127.

Weidenreich, F. 1945. The brachycephalization of recent mankind. *Southwestern J. Anthropol.* 1(1): 1–54.

Weil, A. 1929. Measurements of cerebral and cerebellar surfaces. Comparative studies of the surfaces of endocranial casts of man, prehistoric men, and anthropoid apes. *Am. J. Phys. Anthropol.* 13: 69–90.

Weinert, H., Brauermeister, H., and Romaine, A. 1939. *Africanthropus njarasensis*. Beschreibung und phylotische. Einordnung des ersten Affenmenschen aus Ostafrica. *Ats. f. Morph. v. Anthrop*. 38: 252–308.

White, T.D., Asfaw, B., Degusta, D., Gilbert, H., Richards, G.D., Suwa, G., and Howell, F.C. 2003. Pleistocene *Homo sapiens* from Middle Awash, Ethiopia. *Nature* 423: 742–747.

Witelson, S.F. 1983. Bumps on the brain: right-left anatomic asymmetry as a key to functional lateralization. In S.J. Segalowitz (ed.) *Language Functions and Brain Organization*. pp. 117–144. New York: Academic Press.

Wood, B.A. and Richmond, B.G. 2000. Human evolution: taxonomy and Paleobiology. *J. Anat.* 196: 19–60.

Zihlman, A. 1982. Sexual dimorphism in *Homo erectus. 1er Congres Internationale Paleontologie Humaine,* Nice. J. Piveteau, H. deLumley, and M. deLumley (eds). pp. 949–970.

4 Revisiting Physical Anthropology in India

R.P. Srivastava

Physical Anthropology and its modern usage Biological Anthropology are interchangeably used. In this paper, I use Physical Anthropology because of historical reason as the term is firmly rooted in our literature. India refers to the pre- and post-partition era of the subcontinent.

Ridley has given an account of the ways in which "man has studied for centuries the manifestation of life in himself, the lower animals, and the plants" (1947: VII). He has mentioned men of antiquity who had tried to understand the nature of body so that the disease that afflicted it could be cured. Among the great men of 5001 BC, Hippocrates is best known for his observations on the human organism and the rules he laid down for the treatment of sick were of such an original and lasting value that he is regarded as the "Father of Medicine".

The roots of Physical Anthropology may be traced back to the works of classical thinkers and in the Ayurveda (Life Science) described in the literary Vedic texts. The ancient Indians were interested in human life. They were the pioneers in Ayurveda and practised medicine. The nature of methodic exposition of variation in men by these experts has been elaborated by Kurulkar (1958). For example, they laid down the rules to differentiate Atura (patient) from Prakriti (normalcy). The suitability of man and woman to each other for mating was also examined. They devised methods of measuring various parts of the body by fingers for assessing whether a person has correct proportionate body or not. Interestingly, the Ayurvedic practice in medicine has been in vogue from ancient Vedic period to modern times.

India has a long history of using the fingerprints for identification. Folklore in India according to Katakkar (1992) has certain way of reading fingerprints as indications of attributes or traits. The finger pattern types are known as Chakra (whorl), Sankh (both radial and unlar loops) and Sakti (composite). When loop is found on one finger, the subject is happy, if whorl on two fingers is there, the person is wealthy and a composite on one finger means happiness. These are some of the examples though there are so many combinations of the pattern types and their occurrence on the number of fingers.

We do not know exactly who these ancient people were who have left their cultural achievements. The story will be clear if we know who these people were and where they came from. The skeletal remains available belong mainly to the Indus valley civilization dating back to 2500 to 1500 BC. The important sites from where the skeletons were recovered are the cities of Mohenjodaro and Harappa (Sindh) in Pakistan, Lothal (Gujarat) and Kalibangan (Rajasthan) in India. There are other skeletal remains which have been dated as pre- and post-Harappa's. The pre-Harappa relics are from Pakistan (Nal in Baluchistan) and the post-Harappa relics are from India (Nevasa, Chandoli, and Inamgaon in Maharashtra and Lothal in Gujarat). These specimens are distinct from those of Mohenjodaro and Harappa in their anatomical details.

Let us first consider the views of certain eminent scholars and then examine the conclusions arrived at by anthropologists in their investigation of Indus valley skeletal material. According to Mitra (1949), the people who inhabited the prosperous cities of Mohenjodaro and Harappa were of four different ethnic types. He also suggested that many groups of people entering India from outside intermingled when people of various stocks occupied a common land in India. Munshi (1974) hypothesized that people having varied speech forms migrated into the Indian subcontinent. Among the early migrants were the Negrito from Africa, the Proto-Australoids, early Mediterranean people, and the Mongoloids found in the northern and eastern parts of the country. The later Mediterraneans identified with the early Dravidians, the Alpines considered as the pre-Vedic, Aryan people, and later the Vedic Aryan considered as the Nordics constituted the Indian people. The Aryo-Dravidian synthesis has also been elaborated by Munshi. Before the arrival of the Aryans, the migrating tribes of early Dravidians occupied the greater parts of western and northern India. There is sufficient evidence of marked Dravidization of Aryan languages and infusion of Dravidian elements in Aryan way of life, religion, ritual, and thoughts. The Aryo-Dravidian synthesis took place in the plains of North India between 1200 BC and 700 BC. We refer to Kennedy, who has rejected such an Aryo-Dravidian synthesis. I quote his viewpoint: "We know nothing about the skeletal biology of the bringers of Indo-European languages to South Asia, least of all the early composers of the Vedic accounts. Nor is there any sound evidence to indicate that the so-called Aryan invasion was coincident with the contributory to the decline of the Harappan civilization. The question of a massacre at Mohenjodoro is questionable, and to ascribe the fate of its victims to the Aryans would be unfounded" (1984: 104). Wheeler (cited in Kennedy) attributed the decline and fall of the Indus valley cities to the Aryans. According to Munshi (1974), the military conquests of the Aryans were of minor significance but the cultural vitality was more important. The latter part of his observation on the Indus valley civilization cannot be ignored.

It is necessary before proceeding further to inform the readers that an erudite and visionary Vice-Chancellor of Calcutta University, the late Sir Ausutosh Mukherjee saw the importance of anthropology and took initiative in introducing it in the curriculum of the post-graduate studies leading to MA and M.Sc. degrees. In the history of Indian universities, it was for the first time that anthropology got recognition as an academic discipline in Calcutta University in 1918. Upadhaya informs us that S.C. Roy delivered a lecture in Patna University on the principles and methods of Physical Anthropology, a year before an independent department of anthropology was started in Calcutta University in 1920 for teaching all major branches of anthropology (Wikipedia). The first generation products of this department of anthropology established their own departments in other universities and produced a second generation of anthropologists. The late Professor D.N. Majumdar, a product of the Department of Anthropology, Calcutta University and Ph.D. from Cambridge University, was intellectually the most prominent anthropologist for over 30 years. Majumdar not only established the Department of Anthropology at Lucknow University but also founded the quarterly journal, *The Eastern Anthropologist* of which he was the editor until his death.

The late Professor P.C. Biswas another product of Calcutta University, Ph.D. Berlin, established the Department of Anthropology at Delhi University. These professors trained their students as social and physical anthropologists, and many of them made their mark in whatever field of specialization they worked.

Before the teaching of anthropology began in Calcutta University, it was Sir Herbert Risley (1915) who, while supervising the Census of India in 1900, made the first systematic attempt to racially classify the people of the Indian subcontinent. He identified seven physical types and described them by linguistic terms only six terms as Turko-Iranian, Indo-Aryan, Seythio-Dravidian, Mongolo-Dravadian, Mongoloid and Dravidian.

After Risley, Guha (1931) proposed a different racial classification. Since he had already conducted an athropometric study of the Mohenjodaro skeletons and identified four racial elements which were Mediterranean, Proto-Australoid, Alpine and Mongoloid, it became convenient to include these racial types in his scheme of racial classification based upon the anthropometric measurements of approximately 2,500 living individuals.

Majumdar's anthropometric surveys of three states of India, namely, Uttar Pradesh (1949), Gujarat (1950), and Bengal (1958), are most important from the viewpoint of data analysis and the interpretation of the results.

In his anthropometric survey of Uttar Pradesh, Majumdar in collaboration with Mahalanobis, made use of the most sophisticated statistical technique known as D^2 statistics developed by Mahalanobis. The anthropometric measurements of 22 populations that he took into account were found to form clusters depending upon their similarities. The overall picture that emerged pointed to a gradual lowering of the racial status from the Brahmins to the aboriginal tribes and the order to social precedence corresponded with the degree of racial purity or hybridization. For example, the Brahmins who occupy the highest position in the Hindu society have the largest build up of face and body with tall stature. As one moves down from top to the bottom of the picture there is a steady decrease both in social status and physical size.

Racial status of castes and tribes in Uttar Pradesh that Majumdar figured out are: (1) A long-headed and largely fine-nosed type represented by the Brahmins, Khatris and the Kshatriyas and also in some ways the upper and pedigreed Muslims. This group according to Majumdar is the Indo-Mediterranean (2) A long-headed, flat-nosed type represented by the tribal elements in Mirzapur district. This group is called as Proto-Australoid. (3) An intermediate-headed type with medium to short nose, represented by the Tharus and other Mongoloid elements in the cis-Himalayan region. This type is the Mongoloid.

In the Gujarat racial survey, Majumdar (1950) collected, anthropometric data on 24 populations including tribals (Bhils) and non-tribals. He found that the Bhils and artisan castes were long-headed and the non-tribal populations Bhatias, Rabaris, Parsis, and Miana Muslims were medium-headed.

Karve (1951) also collected anthropometric data from Karnataka, the neighbouring state of Maharashtra. She observed that the primitive people of this state were on the border of long and medium heads (dolicecaphaly and mesocephaly). Besides the primitive people, there is a small but significant group of semi-primitive agriculturists who are long-headed (dolicocephalic). This group represents an ancient agricultural population of Maharashtra which had the possession of the land before another medium-headed (mesocephalic) group came in represented by the Brahmins, Marathas, and the professional castes. Regionally, these medium-headed people connect themselves with the caste groups of Gujarat. Hence, Karve opined that the medium-headed people came into India via north-west and settled in the broad and fertile valley of central Maharashtra and further penetrated deep into Karnataka and the Telangana region of Andhra Pradesh.

Majumdar's Bengal racial survey included only 12 social groups for D^2 statistical treatment. The striking features of this investigation were clustering of the groups more on the basis of the region to which they belonged rather than on caste or religion.

Majumdar (1961) in his analysis of racial composition of the people of India suggested that the dominant racial type in South India is represented by long-headed, narrow nosed, stock of medium stature and brownish complexion which has mixed in varying proportions with the tribal groups on one hand and on the other hand, a broad-headed, narrow-nosed element whose influence is well marked among the people of Karnataka and Tamil Nadu. This broad-headed element known as Haddon's Alpine type must have filtered into India in prehistoric times and is widespread throughout the central belt of India from Gujarat through Madhya Pradesh to Bengal. To the north, this type has mixed with fair, long-headed type named by Risley as the Indo-Aryans. To the east, the broad-headed type has been assimilated by the Mongoloid tribal groups. Majumdar has also observed that there is a gradual lowering of stature and physical build as one proceeds from Punjab to the peninsular India.

Sarkar (1954) in his racial classification of the people of India failed to avoid the use of linguistic terms such as Dravadian, Indo-Aryan, and geographical boundaries. The plus point in his classification is the absence of Guha's Negrito element about which there is certainly no positive evidence. The Autraloids known by different names, such as Proto-Australoid, Pre-Dravadian, Nishada and Veddid were at one time according to Sarkar widely distributed throughout India. They continue the earlier subtraction of Indian population and their make up is present among all the castes of India with its maximum concentration in the lower castes. The Australoids are short statured people with dark complexion. They have long heads, narrow noses, and wavy hair. Certain tribes of South India such as Urali, Kanikar, Malapantaran, Paniyan and Kadar have preserved the original form of Australoid.

The above-mentioned racial classifications are not only of the traditional/typological type but are arbitrary and confusing. Racial analysis of typological type is out of date in Physical Anthropology.

Guha asserts that the Negritos were the first comers and true autochthons. His observation of sporadic occurrence of woolly hair among the Kadars of South India does not hold good. Investigations on anthropometrics, blood group and deromatoglyphics of the Kadars do not indicate any Negrito trait. It is suggested that the woolly hair observed by Guha might have developed from wavy hair as a result of mutation (Sarkar, 1954; Majumdar, 1961).

In fact, the Negrito element is characteristic of the Andaman Islands where the hunting and food gathering tribes (the great Andamanese, Jaravas, Sentinelese, and the Onges) are of the Pygmoid stock, but they have no share in the making of the Indian populations. These Negritos definitely belong to the Oceanic Negroid stock. The origin of this stock is difficult to establish. It is possible that the African and Oceanic Negroids evolved from a common ancestor, but it is equally likely that they were convergently evolved as a result of similar selective forces in the two regions.

Physical anthropologists in India were not oblivious of the developments that were taking place in the West. The decline of interest in the racial analysis among the physical anthropologists was due to collapse of typological approach and the rise of population genetics, a subject founded by three great theoretical biologists, Fisher (1930), Wright (1931), and Haldane (1932). At present, more than two dozen universities in India have full-fledged and subsidiary department of anthropology. The number of anthropologists produced by the Indian universities is second to the USA. The contribution made by Indian physical anthropologists is immense. The largest numbers of physical anthropologists are employed in the Anthropology Survey of India (ASI), Government of India.

The ASI was set up in 1945 at Banaras (Varanasi) by the Government of India with Dr B.S. Guha as its first Director. Later it was shifted to Calcutta in 1948. In fact, before the creation of ASI, anthropology was a section of Zoological Survey of India.

Initially, ASI was required to collect scientific information about the people of India for the use of researchers, planners, and administrators. Now, with its head office in Calcutta, the survey has seven regional offices and one sub-regional office at Port Blair, Shillong, Nagpur, Mysore, Dehradun, Jagdalpur (sub-regional office), Udaipur and Calcutta.

The ASI continued to study the tribal and non-tribal people of India with emphasis on basic research as well as applied aspect in social-cultural and physical anthropology. The skeletons from Mohenjodaro and Harappa have been anthropometrically studied and described by several investigators. Mention may be made of Guha (1931) who identified four racial elements at Mohenjodaro, which he called Mediterranean, Proto-Australoid, Alpine and Mongoloid. Sen (1964–65) has questioned the method of racial analysis from skeletal remains. He pointed out that the interpretation of the skeletal material has been made on erroneous and outmoded concept of race. His emphasis was on genetic definition of race. Dobzhansky and Dunn (1958) defined race as "populations that differ in the relative frequencies of gene alleles or of chromosomes structure". This definition presupposes that a population consists of intermarrying individuals as they are but partially reproductively isolated from other such populations. Hence, the criterion of reproductive isolation enables us to make a clear-cut distinction between species and its sub-species. Species is a close unit and the sub-species partially so. It also means that there is an intergradation between two species due to gene flow.

In the light of the above-mentioned facts, Sen opined that the skeletal material of Harappan date from each site represented a homogeneous population with regards at least to head shape, nose shape, and stature. This would mean that of whatever original compositions the populations were at these sites each belonged to a single biological population and not recognizable as belonging to several races with distinctive characteristics. This population was broad nosed, tall, and long-headed in the Punjab and Sindh, with somewhat larger head in Gujarat. Kennedy (1984) also does not agree with Guha's observation because his racial identification is based on a limited number of cranial features and the variation seen among Indus valley population is not beyond the normal range of variation for this population as a whole. Bennet (1969) opines that in physical anthropological studies of human skeleton population, ideal typology must be replaced by population thinking if we want to understand biological attributes of prehistoric populations.

Mario Cappieri (see Kennedy, 1984) in a multivariate analysis of Harrappan skeletal data along with those from archaeological sites outside of South Asia but of comparable antiquity, found closest affinities between the people of Harappan civilization and the ancient populations of the Mediterranean basin and West Asia rather than with the peoples of peninsular India and the land lying to the east of Brahmaputra. Cappieri's observations are in conformity with that of Kennedy that the Mohenjodaro-Harappan population represented an eastern sector of a Mediterranean Western-Asiatic phenotypic pattern and that the ancient Harrappans are not markedly different in their skeletal details from the present-day populations of northwestern India and Pakistan who again share a lot of characteristics with both the ancient and living peoples of Afghanistan, Iran, and the lands bordering the Aegean and the Mediterranean.

The natural geography of the Indian subcontinent is such that migration into this region might have taken place both along the sea and over land. There have been series of migrations from the North-East, Tibet, South China and Southeast Asia. Of these, the migration from North-West was the most significant for the inhabitants of the Indian subcontinent which are predominantly Mediterranean in origin. Their phenotypic and Rh blood group pictures show a remarkable resemblance to that found in the Mediterranean area. The people of the Indian subcontinent are all of Mediterranean affinities baring the Veddoids called Proto-Australoids and the Mongoloids. Within the group of Mediterranean we may recognize two groups of people. The southern one represented by dark-skinned population speaking Dravadian languages and the larger northern one represented by lighter skinned, typically Mediterranean populations speaking Indo-Aryan languages (Mourant, 1954; Sonia, 1963).

As stated earlier, physical anthropologists in India were not oblivious of the developments that were taking place in the West. The decline of interest in the racial classification was due to collapse of typological approach and the rise of population genetics. Population was recognized as the unit of evolution and not the individual. Coon, Garn and Birdsell (1950) questioned the non-adaptive nature of racial traits that were used for studying human variation. Newer traits such as blood groups with known genetic basis were used by Boyd (1950) with the assumption that they are non-adaptive. But this approach was short-lived as the data available in Darrow's (1930), Levine and Stetson's (1938) works were already there to show that blood groups were not non-adaptive which was well established later by the investigation of Aird (1953). In spite of these findings, there was still a feeling among the physical anthropologists that genes could serve as race markers. However, after the discovery of blood groups, genetic variants such as serum proteins, enzymes, and immunogenetic variables were discovered one after another in the mid century. The mapping of the gene frequencies of various genetic variants enables physical anthropologists to identify the relationship of some genes with particular variables of the environment and clues to the adaptive significance of particular variants (Roberts, 1970). "The discovery of the selective advantage for sickle cell heterozygotes was one of the major factors in the rediscovery of selection in the 1950's as an important evolutionary force" (Livingstone, 1980).

Dermatoglyphics considered as a non-adaptive trait was used in studying the population variation and as such finger and palm prints of over 17,000 individuals belonging to the tribes from different regions of India have been analysed for ridge configuration and ridge counts by ASI. The ASI has completed two large-scale anthropometric surveys (1961–72) and (1972–77). The details are published in various journals and seminar publications.

Serogenetic survey by ASI based on ABO, MNS and Rh blood groups for more than over 100 population groups belonging to different castes, tribes, and other communities from Kashmir to Kerala and from Nagaland to Kutch, throw light on the origin and movement of population and on the process of genetic adaptation in response to varying environmental stresses. The determination of sickle cell trait in about 17,000 individuals is of special importance.

ASI also offers fellowships and participates in collaborative research. There are 48 fellowships offered to visiting fellows, post-doctoral fellows, senior fellows and junior fellows who work in ASI and in departments of anthropology and cognate disciplines in Indian universities. The most important and extensive investigation, *The People of India Project* was undertaken by the ASI under the supervision of its Director General Dr S.K. Singh (1994). In this project, the ASI assigned the entire Indian population to 4,635 communities spread over 32 states and union territories. The database has been published into several volumes, which provides information on cultural, ecological, and economical traits of all the communities.

Majumdar's blood group investigations are not confined only to various castes and tribes of United Provinces (now Uttar Pradesh) but also to Gujarat. The blood group gene frequencies can be seen in his publications (Majumdar, 1947; Majumdar and Bahadur, 1952).

The various castes of Uttar Pradesh (UP) are found to vary with respect to the incidence of A blood group and there is a gradual tapering of A percentage from the Brahmin to the artisan castes. The blood group percentages of various castes and tribals of UP have shown that there exists a correlation between blood groups and anthropometric traits. The upper castes show significant distance from the tribals, Muslims, and artisan castes.

Negi (1962) has done a detailed serological investigation amongst the Rajputs of western UP and Bharatpur district of Rajasthan. The ABO blood group gene frequencies (p 20.0% and q 24.0%) found in the Rajputs are very close to that of Majumdar. MN results reveal high preponderance of M gene (69%) over N gene. The basic data and the chromosome frequencies for Rh blood groups indicate Rh negative about 7 per cent and R_1 about 59 per cent.

There is another series of blood group data by Srivastava (1978) from various castes and Tharus, a Mongoloid tribe of UP. The two series of data from UP (Srivastava's and Majumdar's) were compared to check variation. It was found that in both the series, the frequency of A gene is low and that of B gene high. The only difference between the two series is that Majumdar's data show higher frequency of A gene in all the groups with the exception of Muslims. Srivastava observed that the pooled frequencies of Hindu castes exhibit lower frequency of A gene and higher frequency of B and O genes as compared to the pooled gene frequencies of Muslim castes. The Hindus and Muslims thus can be differentiated from one another as regards the ABO data. Another interesting thing to note is that the higher castes of Muslims are the descendents of immigrants from Turkistan, Persia, Arabia and Afghanistan (Ansari, 1959–60). It will be worthwhile to examine the blood group characteristics of the inhabitants of these countries. According to Mourant (1954), there is preponderance of A gene over B gene. The frequency of A gene varies from 22 to 33 per cent and that of B gene from 11 to 16 per cent. Hence, Ansari's observation seems to be valid on serological grounds. In as much as the present series, all the higher castes of Muslims do show higher frequency of A gene (21%) as compared to the B gene. The incidence of B gene is not as low as one finds in the above mentioned frequencies. The Hindus and Muslims both show high frequency of M gene. There is a fairly uniform distribution of D gene which clusters around 17 per cent. The three groups of Tharus, a Mongoloid population, have similar D gene distribution as the people of the plains. The raised incidence of Rh negative indicates mixture in contradiction to the complete absence of the factor among the Gorkhas and other Mongoloid population to the north of India.

Majumdar's (1950) ABO blood group data from Gujarat suggests that different groups of Kunbi Pattidars, Brahmins and Bhatias form a cluster by themselves in respect of A group frequency. There is less of A and more of B blood group among the Rajmahal Bhils as compared to the Bhils of Panchmahal district, but both are significantly different from that of Macfarlane.

It is interesting to know the distribution of blood groups in populations of Jaunsar Bawar, Chakrata, northern sub-division of Dehradun district, and of Kinnaur, border district of Himachal Pradesh. A brief description of the people of these areas is essential for understanding their relationships based upon blood groups. In Jaunsar Bawar, the society is divided into a less rigid caste system because there are sometimes inter-caste marriages, whereas the Jaunsaris practice a kind of marriage called fraternal polyandry though monogamy and polygamy are not uncommon. The high-caste group includes Khasas, Brahmins, and Rajputs, and the artisan-caste group contains Lohar, Sutar, Bajagi, and the lower caste has Kolta.

In Kinnaur of Himachal Pradesh, the high altitude division Pooh is solely inhabited by the Mongoloid group, who call themselves as Kanet (Rajput) and Koli. The middle division Kalpa is inhabited by mixed populations and the southern-most division Nacchar is inhabited by populations that are mostly non-Mongoloid. The above-mentioned populations were surveyed for ABO blood groups by Negi et al. (1972). Gene A attains a high value in all the Jaunsari population (23–28%) and also in Kinnaur (24–32%) except in Pooh sub-divisions where both Kanet and Koli show 10–11 per cent. Gene B maintains a moderate value in this region barring Pooh where its frequency is as high as 36 per cent in Kanet. Thus, it appears that populations with Indo-Aryan affiliation are characterized by high A gene, the Mongoloids exhibit high B gene and the Proto-Australoid elements possess moderate values of both A and B genes.

The most sensational discovery of historic interest was made in 1956, when a litter of bones was found is Rupkund, a Himalayan glacier lake, at an altitude of approximately 18,000 feet in the Gurhwal district of UP (now in Uttarakhand).

Mr Mahesh Chandra and Mr Natheyal, Forest Officer brought down a few skulls, jaw bones, long bones, and a number of artefacts, shoes, beads and bangles. All these were handed to the Department of Anthropology, Lucknow University. A team of experts from different disciplines under the leadership of Professor D.N. Mujumdar reached the lake in the month of September of the same year and brought down a rich collection of bones, artefacts, specimens of hair, textiles, ornaments, utensils, tent equipment, pegs, nails and knife blade.

Anthropometric measurements, blood groups from dried flesh, fluorine content of the bones and density of the bones were completed in the laboratory of the department. Radiocarbon dating was done in the Phoenix Memorial Radiocarbon Laboratory, University of Michigan. The bones were found to be 650 years old with a range of 150 years on either side.

On the basis of the results of various tests, Majumdar concluded that the Rupkund victims were of a ethnically homogeneous stock similar to the people of north-western India, 300 miles away. The results were also examined by Dr G.M. Mourant, an outstanding authority on anthropometry and craniometry. He agreed with Majumdar's conclusion.

There were many theories which could not be supported by the scientific evidence. Majumdar did accept that he could not propose any plausible hypothesis, but believed that people were from north-west parts of India and they may have been an invading army destroyed by natural calamities. The absence of arms is not important as in every battle the reconnoitering army picks up arms. The absence of women is difficult to explain. Even the armies carry women along (unpublished).

Srivastava et al. (1979) carried out the first serological survey (ABO) in a sample of 1,399 individuals belonging to various sub-caste populations of Brahmins and Lingayats in Karnataka where the people speak Kannada, one of the Dravidian languages of South India. Among the Brahmin groups (Vaishnavas, Shaivites and Havayaks), the gene frequencies of both A and B clusters were around 18 per cent. Among the Lingayat groups, the B gene was higher than the A gene.

Sen (1960) examined a sample of Bengalees comprising Brahmins, Kayasthas, Vaidyas, and a small number of lower castes for A_1 A_2 B O, MN, Rh, Kell, and P blood groups, the secretor factor, and for haemoglobin variants. He found that the three castes were similar in blood group frequencies and the secretor factor was low in the Vaidyas. Haemoglobin E and D were found to be present in low frequencies. While assessing the racial composition of Bengalees, Sen (1962) observed that when the B gene is compared with other Indian populations, it is found that the Bengalees with their higher B show a marked difference from the populations of Madhya Pradesh, Maharashtra, Madras and Kerala. There are other serological systems in which the difference is further extended into P and secretor factor. Sen made further observation by taking into account the haemoglobin variants. It is known that Gujaratis, Punjabis, and Bengalees have HbD in more or less equal low incidence. It appears that the simultaneous presence of HbD and HbE in Bengalees would indicate that the direction must have been from the west to the east rather than from Bengal to the west, otherwise Sikhs would have also shown HbE which is absent among them.

Bhalla's survey of literature from 1936 to 1966 is related to the gene frequency distribution of ABO, MNSs and Rh-Hr blood groups. He considers the salient features under five major zones, namely western, northern, eastern, central and southern on the basis of ethno-geographical variations in the Indian subcontinent. There is a good deal of variation (A 20% and B 60%) in the western zone. The northern zone is characterized by high frequency of B than A. In the eastern zone, the tendency of higher A than B is met with greater regularity. The central zone is mainly an area inhabited by a large number of tribes. They exhibit higher frequency of B than A. The southern zone is represented by consistently higher incidence of B than A. High frequency of O is seen in Ceylon and the Andaman and Nicobar. The Indian subcontinent is characterized by high frequency of gene M than much lower frequency of gene N. As regards the distribution of Rh factor, there is preponderance of gene D and complete absence of gene D in some tribals of Chhota Nagpur. Extremely high frequency of chromosome R_1 (CDe) is an important feature of the Indian subcontinent.

Sanghvi (1953) was the first to measure genetic distances between five endogamous groups in Bombay on the basis of blood groups, PTC and colour blindness data. None of the pairs of the groups show significant difference. Karve and Malhotra (1968) have tested the hypothesis that sub-castes are probably the result not of fission but of lack of fusion. They have taken into account many anthropometric measurements, ABO, MN, and Rh blood groups, colour blindness, hypertrichosis, and hand-clasping of eight endogamous groups all calling themselves Brahmins in Maharashtra. They used various statistical methods developed by Sanghvi to assess quantitatively the biological affinities between these groups. On the basis of the statistical results, they found indications of independent origin for five of the eight groups. They emphasize that the data confirm a sociological hypothesis such as this one, yet there is nothing in the data to contradict it. Karve and Malhotra's approach to the problem of sub-caste origin has been appreciated, criticized, and suggestions offered. For details see CA treatment, where the comments by experts and the reply by the author are published along with the article.

Gulati (1971) investigated nine endogamous groups of Kumbhars (Potters) in Maharashtra to test Karve's hypothesis. His results partially support the hypothesis as some of the Kumbhar groups show morphological as well as genetic (blood groups) affinities whereas others are quite independent genetical categories. I think that conclusive evidence in support of the hypothesis maybe available from the comprehensive DNA study if somebody ventures to investigate.

Basu (1969) conducted a genetical study of the Pahiras, a small primitive population divided into three partially endogamous sub-populations. He has named the three Pahiras sub-populations as northern Pahiras (NP), who inhabit the foothills of Ayodhya hills in the Purulia district of West Bengal and the southern Pahiras (SP1 and SP2) who inhabit Dalma hills in the districts of Singhum and Ranchi, Bihar, now Jharkhand. The data of ABO blood groups show highest frequency of gene A both in NP (37.26%) and SP1 (31.05%) while in the SP2, the highest frequency is of B gene (32.99%). The difference between NP and SP2, and between SP1 and SP2 are statistically significant. In taste blindness, NP is characterized by high frequency of tasters (58.45%) than non-tasters, while the reverse situation occurs in SP1 and SP2. As in blood groups, NP is closer to SP1 than to SP2 in taste blindness. What is important in this study is that demographic information on breeding structure suggests that allele frequencies throughout a population are not homogeneous.

Sen and Pal (1968) for the first time in India investigated the role of genetic drift as an evolutionary factor. Treating the demographic data from a village consisting of several endogamous groups near Calcutta, they observed that two castes, namely, Dhibars and Kayasthas show a low index of isolation signifying that random genetic drift affected the gene frequencies of these two castes.

Balkrishnan (1978) in his study of genetic distances (a measure of difference between gene pool) among some populations of the Indian subcontinent found that the populations do cluster and follow certain patterns. He has taken into account the results of ABO and Rh blood groups (with four and five antisera) obtained from 102 populations spread from the Northwest Frontier Province (in Pakistan) to Assam, Kashmir, Bhutan and Sri Lanka. Twenty-three populations from the northwest, 45 from the west, 20 from the south, and 14 from the northeast were included. There were 14 Brahmins, 40 other varnas (traders, agriculturists and artisans), seven Harijans, 28 tribals, eight Muslims, three Christians along with Parsis and Jews. The magnitude of genetic distance was utilized for clustering of the populations. The northwest was the most homogeneous region and northeast was the least. The heterogeneity in the west and south showed a dramatic reduction on exclusion of the tribals and no change was found even if the tribals were included. It is seen that the clustering between the tribals and non-tribals is quite clear with few exceptions. The non-tribal cluster of 39 populations contains all the 23 populations from the northwest, five Maharashtrian Brahmins, and all the three Lohanas. It also includes three tribals – the Toda, Irula and Vaghri. According to Balakrishnan, the positioning of these tribals into the non-tribal cluster could be a product of interaction between genetic composition and selection through extreme cold. Since the Irulas have the closest distance with the Rajput of Kumaon (UP now Uttarakhand), and the Todas also have similar distance with populations from this region, Balakrishnan speculates that these two tribes maybe the remnants of earlier immigrants from this region into the south and the Vaghri is a segment left off during the migration. The Todas are not only pastoral but also polyandrous, a custom also seen in the cis-Himalayan region. The Irulas do not eat the flesh of cattle or buffaloes and according to Kirk et al. (1962) they are different from the Australoid Veddas with whom they have traditionally been associated.

The other non-tribal cluster of 35 populations contains 12 out of 15 non-tribal South Indian population. The tribal cluster is predominantly Australoid but includes some of the non-tribal populations such as Maratha, Bhangi, and Mahar of Maharashtra, Rajbansi and Muslims of Bengal. Balakrishnan opines that all of them probably have a high Australoid content. The other tribal cluster is predominantly Mongoloid though there are certain exceptions. Balakrishnan tentatively names four clusters as Causcasoid (Aryan), Caucasoid (Dravadian), Australoid and Mongoloid.

The picture drawn for India, Pakistan and Bangladesh also applies to Nepal, Bhutan, Sri Lanka and Maldives. The people of Nepal and Bhutan are basically of Mongoloid stock. The Sinhalese of Sri Lanka are no different from the Tamils of South India. Kirt et al. (1962) in their genetic investigation found the similarity between the Sinhalese and Tamils in Sri Lanka and the Tamils in South India more striking in contrast to the relationship between the Veddas of Sri Lanka and the Veddoid Irula of South India. Kirk and others opine that if the Veddas of Sri Lanka and the Veddoid Irulas came originally from the same stock their long isolation led to a marked divergence in gene frequency.

There is a lot of controversy as regards the origin of Dravidians and Aryans. According to one school of thought, the advent of the Aryan speakers in the northwest of the Indian subcontinent has been dated to about 1500 BC. They were barbaric nomads and presumed to have destroyed the cities of Indus valley civilization. The origin of the speakers of the Dravidian languages is also obscure. However, archaeological evidence indicates that about the middle of the first millennium BC, an intrusive iron using megalith-building civilization suddenly appeared in the Deccan and southern India. This civilization has been attributed tentatively to Dravidian-speaking invaders. The only Dravidian speaking outside India are the Brahuis of Baluchistan who may provide some clue to solve the Dravidian mystery.

We have seen in the foregoing discussion that conventional genetic markers such as blood groups were used for the analysis of human variation. But with the discovery of nuclear and mitochondrial DNA polymorphism, the interpretation of genetic variation in human populations has become more lively and reliable.

Negi (1963) and Kumar (1965) have carried out extensive surveys on the sickle cell trait in South Bastar, north Bilaspur district, Ratlam, and adjacent districts of Madhya Pradesh. Negi found that the frequency of sickle cell trait varies from 19.39 to 21.26 per cent among the three groups of Gonds tribe of Bastar, whereas Kumar observed only 5.9 per cent among the Bhils of Ratlam. In an earlier investigation, Kumar (1960) did report a frequency of 12.06 per cent in Bhils and 28.05 per cent in another Bhilara tribe. The distribution of sickle cell trait in South Indian tribes has been reported by Malleswara Rao et al. (1979) and

Sudhakar et al. (1980). Only two cases of sicklers were found in Savara and Jatapu of Srikakulam district of Andhra Pradesh, whereas Babu observed 24.24 per cent in Koya Dora of coastal region. Though there are many investigations in other tribal and non-tribal populations in India, the maximum incidence has been found in the tribals of Madhya Pradesh and Andhra Pradesh. This trait has also been found in some lower castes but not in the upper castes. The presence of this trait in homozygous condition, though lethal, offers selective advantage to heterozygous individuals rather than the homozygous normals in endemic malarious area. How far this observation holds good in the case of India as the trait is mainly found among the tribals in the western and central India for these regions do not correspond to any specific environment. The sickle gene must have arisen by mutation.

The study of haemoglobin variant was first undertaken by Sen (1960) among Bengalee caste populations settled in England. Haemoglobin E and D were in low frequency. His conclusions have been mentioned along with the analysis of blood groups in Bengal. Tiwari (1960) for the first time also worked out the frequency of haptoglobins among the upper castes of Bengalees settled in England. The frequency of Hp^1 gene was found to be 20 per cent. The frequency of this gene is fairly low in India as compared to western Europeans. South Indian tribes have shown some of the lowest values (0.7%) with the exception of Todas in whom the frequency reaches maximum of 37 per cent. Tiwari (1961) reported in another study that among the Punjabis there was 20.94 per cent of Hp^1 gene and the only transferring Tfc. Haptoglobin types have been found out in some Indian populations by many investigators. Kaur and Chattopadhyay (1998) tabulated the distribution of haptoglobin types in 14 populations of Punjab and found Hp2-2 haptoglobin types in 14 populations. Hp2-2 was found to be most common with a frequency of 64.58 per cent in Jat-Sikhs and 69.30 per cent in Baniyas. Hp1-1 was found in a very low frequency of below five per cent. HpO was not found in any population. Livingstone (1985) in his compilation of G6PD frequencies worldwide, observed that this mutant has very low incidence and appears to offer some selective advantage in endemic malarious regions. In India, its lowest frequency of 7 per cent is seen in Assam and as high as 12 per cent in the tribals of Andhra Pradesh.

There are many surveys on other protein polymorphism in Andhra Pradesh. Walter et al. (1978) investigated the serum protein polymorphism in three tribals and found considerable differences in the distribution of gene frequencies among them. Busi et al. (1979) conducted biochemical investigation on some red cell enzyme group system. Of the six systems investigated, only acid phosphatase and 6-phosphogluconate dehydrogenase showed polymorphism. In the red cell, acid phosphatase p^c gene is 0.03 which can be called high as it is absent in most of the Indian populations. The incidence of p^b is very high (0.71) a characteristic of Indian populations. The incidence of PGD^c allele is rare and is in conformity with the data found in other Indian populations.

Das et al. (1980) have investigated the distribution of HbE in six Mongoloid populations of Assam in which the gene occurs in very high percentage ranging from 44.7 to 54.9 per cent. Walter (1986) compiled information on four polymorphic systems – Tf, Ge subtypes, Gm, and Km from the published literature on Indian populations. He concluded that the caste groups and tribals of northern, southern and western India are genetically closer to the West-Asian populations, whereas the east Indian populations are more similar to Mongoloid. Thus, it is quite clear that HbE is most frequent in the eastern region of the country. HbE occurs frequently in combination of ß thalassaemia disease. Beta thalassemia major has been found in most of the states of eastern India. Cases of this trait have been observed in UP, Sindh, and Punjab. The preponderance of HbE thalassemia over ß thalassemia may be attributed to the relatively milder effects of the former, which are seldom fatal (Sharma, 1983). Mukherjee et al. (1985) designed a survey of 12 polymorphic systems –ABO, Gm, Km, Hp, cp, Tf, Alb, Hb, aP, EsD, AK, and PGM, in 10 endogamous groups of West Bengal. The sample included high caste, scheduled castes, and tribes. They found considerable genetic variability among the 10 groups. However, in spite of inter-group variation, the statistical analysis of the data showed clustering of the populations. The clusters reflect the existence of three racial groups, the Caucasoid, Mongoloid, and Proto-Australoid.

In a jointly conducted study, the researchers from Andhra University and the University of Utah, USA used the latest molecular tests to find out the genetic makeup of certain castes in Andhra Pradesh. The researchers – Lynn B. Jorde from USA, B. Bhaskar Rao, and J.M. Naidu from Andhra University have suggested that DNA of the people in upper castes have similarities with Caucasians. The genetic distance between the upper and lower castes is one and a half times greater than between upper and middle or middle and lower castes (*The Indian Express*, November 9, 1988).

Michael Bumshad from University of Utah, USA, has in collaboration with Andhra University and the University of Madras, shown that each caste's mitochondria which is derived from the mother only, has greater similarity to Asians than to Europeans but the upper castes show less similarity than the lower castes. Conversely, Y chromosome, derived from the father only, shows each caste is similar to the Europeans with the upper castes being most similar. This has led Bumshad and his associates to hypothesize that more Eurasian males migrated to India than did Eurasian females. They have further examined 40 additional genes that are inherited from father and mother. All of these data strongly supported that upper castes have greater genetic similarity with the Europeans than the lower castes. In fact, the genetic distance is closest between Europeans and Brahmins followed by Kshattryas and Vaishnavas (*The Times of India*, May 21, 2001). Since the study is confined to a particular region, no definite conclusion is possible.

Watkins et al. (1999) found that mitochondria mtDNA9-bp deletion has arisen in some Indian tribal populations. The results indicate varying genetic affinities of South Indian tribes to continental populations. Watkins et al. (2005) analysed 45 unlined STR loci in 151 individuals from tribal groups distributed over several states of India (Andhra Pradesh, Karnataka, Tamil Nadu, Kerala, Orissa, and Madhya Pradesh). For comparative analysis, they included eight castes from Andhra Pradesh and 18 populations from Africa, Europe, and East Asia which were already investigated by others. The results of their study indicated that the tribal Indians are more closely related to caste Indians than to other major groups.

Very interesting results of an important collaborative genomic study have been published under the title "Aryan-Dravidian divide a myth study" (*The Times of India*, September 26, 2009). Lalit Singh, former Director, Center for Cellular and Molecular Biology (CCMB), Hyderabad and a senior scientist of the Centre Thangarajan collaborated in the study with researchers at Harvard Medical School of Public Health and the Broad Institute of Harvard, and MIT.

The team analysed 5,00,000 genetic markers across the genomes of 132 individuals from 25 diverse groups from 13 states. The individuals were from six language families representing upper, lower, and tribal groups.

The finding of this study led the researchers to conclude that the castes grew out of tribal-like organizations during the formation of the Indian castes. It is not possible to distinguish between castes and tribes as their genetics are not different.

One of the researchers, Thangarajan made the following observations:

1. The earliest settlement took place 65,000 years ago in the Andamans and ancient South India around the same time.
2. At a later stage, 40,000 years ago, the ancient North Indians emerged.
3. At the same point of time the ancient North Indians and the ancient South Indians mixed giving rise to a different set of population, the population that exists today. Since the Indian populations are genetically related, the *Aryan Dravidian divide is a myth.*

The discovery of markers known as mtDNA variants made the distinguishing of populations from different regions. M20 is one of the variants of mtDNA. According to Wells (2003) M20 indicates the first major wave of migration into India from the Middle East around 30,000 years ago. It is found at highest frequency (over 50%) in the population of south who speak Dravidian languages – a language family unrelated to Indo-European. M20 is found only sporadically outside India. Wells believes that M20 is an Indian marker. Wells tried to complete the analysis by pointing out the presence of M17 a derivative of M173 which is consistent with a European origin. The origin, distribution, and age of M17 strongly suggest that it was spread by Kargan people in their expansion across the Eurasian Steppe. M17 in India is found at high frequency in people who speak Indo-European languages, but the Dravidian speakers exhibit much lower frequency (10% or less). This evidence suggests that M17 is an Indo-European marker and indicates that there was a massive genetic influx into India from the Steppes within the last 10,000 years. Wells considers the results of Y chromosome from Iran and Middle East as a proof that early Middle Eastern agriculturists did not spread Indo-European languages as they moved into the Indus valley. Thus, it is clear that agriculture pre-dated that of Indo-European languages. Wells' analysis seems to reinforce the archaeologists, argument including that the arrival of Dravidian speakers were the first settlers followed by Indo-European speakers. Regarding the causes of the collapse of the Harappan civilization, recent research suggests indigenous causes such as river changed course and social decay. Wells thinks that the scenario is only speculative (2003).

There are also investigations on the distribution of mtDNA and SRT loci in the tribal and non-tribal populations of north-east region, West Bengal, and Madhya Pradesh (Mukherjee et al., 1999; Chattopadhyay et al., 2001a, 2001b; Dutta et al., 2001, and Mukherjee et al., 2001). These studies were conducted to know the similarities and differences between different groups of Indian people at the level of some DNA markers.

Besides ASI's survey, a large number of caste populations have been studied for their dermatoglyphic traits by different investigators mainly after 1960. Biswas and Choudhari (1938) were the first anthropologists to study the Palmar prints of Oriyas. Srivastava (1963a, 1965a, 1965b) applied univariate *t*-test for the first time in assessing the degree of biological affinity on the basis of pattern intensity index and finger ridge counts among the three endogamous groups of Tharus. Coon (1969) has cited this fact in his book. Srivastava also pointed out that means and standard deviations in the case of Furuhta's and Dankmeijer's indices cannot be worked out because in most of the cases the index becomes either zero or infinite due to the absence of one or the other pattern involved.

Singh and Bhasin (1979) have compiled the dermatoglyphic data from 1936 to 1979 from India. It is observed that by and large Indian populations are characterized by 40 to 55 per cent of whorls. The Mongoloids groups do show lesser number of arches. The pattern intensity index hovers around 14. The commonest Main Line Forumlae 11.9.7. followed by 9.7.5. and 7.5.5. are found in the Indian populations. Mean ridge counts are not available in Singh and Bhasin's compilation. However, mean ridge counts of a dozen populations of UP are available. In Srivastava's (1965a) investigation of Tharus of UP, mean ridge count varies from 116 to 136. Among the three groups of Brahmins it varies from 124 to 148 .(Shukla and Srivastava, 1969). Highest mean ridge count of 161 has been reported by Singh (1961) among the Muslims of UP. Srivastava (1966) in a review of dermatoglyphic research in UP suggested that the mean ridge count as a dermatoglyphic determinant scores over the IPI in statistically differentiating human populations of different ethnic origin.

Srivastava (1959) tested taste sensitivity to phenythiocarbamide among the Brahmins, Vaishas, Kayasthas, Khattris and Muslims of UP. The frequency of non-tasters varies from 24 per cent in Kayasthas to 38 per cent in Muslims. The low frequency of 15.03 per cent of non-tasters is found among the Tharus, a Mongoloid tribe in UP. This low frequency among the Tharus is in conformity with that of Chinese who are also of Mongoloid origin.

There is another series of data collected by Srivastava and Tyagi, from upper castes, lower castes and Muslims of UP (*ASI Bulletin*). All the populations of upper castes appear to be homogeneous. The frequency of non-tasters varies from 24 to 34 per cent. In this series, the frequency of non-tasters is not different from that of the other series.

The Paniyans and Kadar tribes have high incidence of tasters (82–87%). There is also high incidence of tasters among the people of eastern Himalayas and north-eastern India which is similar to the Mongoloid populations. The incidence of tasters varies from 60 to 75 per cent among the Bengalees in West Bengal and a lower frequency from 46 to 50 per cent among the tribes of Orissa. The average incidence of tasters in tribals and non-tribals in Gujarat are 52 and 58 per cent, respectively, whereas in the neighbouring state of Maharashtra, the incidence is generally higher in caste groups than in nomads and tribal group (see Roychoudhury, 1983).

The ability to smell cyanide is another non-antigenetic trait. It has been reported by Kirk and Stenhouse (1953) that some individuals cannot smell cyanide and this deficiency is probably sex-linked Mendelian recessive trait. Srivastava's (1961b) data from UP is the only series from India. The frequency of non-smellers in males is around 17 per cent and in females 5.73 per cent.

Defective colour vision generally known as red-green colour blindness is one of the best-known sex-linked characters in humans. Since the defect is conditioned by a gene located in the X chromosome, it is much more common in males than in females; and the estimated ratio is about 10 colour-blind men for every colour-blind woman. The screening method with the Ishihara chart is quite satisfactory for the detection of colour-vision defects, but an accurate diagnosis involving the sub-types could be better achieved by the use of an anomaloscope. The Indian populations screened so far were exposed to Ishihara chart for the detection of colour-vision defects in the red-green region of the spectrum. Dutta (1966) had analysed the then published and unpublished data consisting of 10,088 males but Srivastava (1983) found many published samples from Uttar Pradesh and Madhya Pradesh unutilized. Since then, many more populations were investigated and the number of males tested for colour-vision deficiency was raised to 17,581. According to Srivastava, the total amount of existing colour-vision data from 17,581 males belong to different castes and tribes. The females have not been included in the present study though the number would be over 4,000. The overall colour-blind frequency for Indian males comes to 3.86 per cent. The population in which the frequency of colour blindness is not only high but twice the average (2.7%) obtained for India are arranged in decreasing order of their frequencies. It is found that the frequency of colour blindness is high in many groups of Brahmins of Maharashtra. Therefore, an attempt was made to see whether the high incidence of defective colour vision is also prevalent among the Brahmins of other parts of the country. The social hierarchy of the Brahmins is the same everywhere as they occupy the highest position in the caste hierarchy throughout the country. The highest frequency of colour blindness is seen to vary from five to seven per cent among the Brahmins of South India (Andhra Pradesh and

Karnataka) followed by those of Western (Maharashtra) and Eastern India (Bengal and Orissa). The lowest frequency two to three per cent is observed among the Brahmins of Central (Madhya Pradesh) and Northern India (Uttar Pradesh).

It is interesting to note that the frequency of colour blindness is lowest in Assam and Tamil Nadu otherwise the incidence is more or less the same among the non-tribal populations of other states. The tribals, in general, exhibit lower incidence of colour blindness. It has been seen that the defective colour vision data do not exhibit any significant inter-state variation. However, the data have been further pooled to see whether any meaningful pattern of regional variation emerges or not. For this purpose, the author divided India into four regions namely, Western India, Northern India, Southern India, and Eastern India. There appears to be uniform distribution of colour blindness in different regions of India.

When the data of the non-tribal group are divided into Hindus and Muslims, it is observed that the range of variability between them is not marked. However, it is interesting to note that the difference in the incidence of defective colour vision between the non-tribal and tribal populations is clearly marked ($X_2 = 29.83$, 1 d.f. $P > 0.01$). Dutta (1966) in his analysis of defective colour vision data from India also found significant difference between the tribal and non-tribal populations. Thus, the results of both studies of Indian data support the selection-relaxation theory proposed by Post (1962). According to him, colour blindness is much rarer in populations of hunters and food gatherers than in those of more advanced culture. This difference is due to the operation of natural selection against colour blindness in the former and by relaxation of the selection pressure in the latter. It seems relevant to point out here that none of the tribal populations tested so far in India belong to the hunting and food gathering category in the true sense, a fact also mentioned by Dutta (1966). However, the merit of the selection-relaxation theory cannot be denied for it does explain the measurable differences in the frequencies of colour blindness between populations of hunters and food gatherers and those of advanced culture. But, the observed distribution of colour blindness in various populations of the world indicates lower frequencies in Asians and

American-Indians as compared to the Europeans, Jews and Russians. It is here that one is reminded of the intrinsic value of the ethnic genetic background of the people so rightly emphasized by Adam et al. (1967).

The study of human growth and development occupies an important place in human biology and biomedical science. In India, nobody has conducted growth studies on genetically dissimilar groups living in similar environment and vice versa. Majumdar did not confine himself only to the racial surveys and the distribution of blood groups in several tribal and non-tribal populations but was also interested in the growth and development of children. He supervised a research project of cross-sectional study of growth of children in UP funded by the Indian Council of Medical Research. A team consisting of physical anthropologists, doctors, statisticians, and social workers collected data on 30,000 children. The consolidated report of all India cross-sectional study on growth and development of children by Majumdar and Biswas published by the Indian Council of Medical Research serves as all India norms for growth.

However, there are few investigations on growth and development by physical anthropologists in certain populations where they have shown the trend of growth in different age groups. In one of the investigations, Biswas and Bhattacharya (1966) observed that pubertal spurt is seen in the age group of 12–14.9 years in Punjab, Rajasthan, as well as in Jammu & Kashmir. The urban group shows an earlier pubertal spurt as compared to the rural for each of the provinces. Singh (1970) in his investigation of various girths and widths of Punjabi-speaking Hindu-Khattri boys found that in all the age groups the mean values of arm girth measurement were lower than those of the American boys. Nirmalananda Reddy (1975) found the rate of growth and maturation in boys of lower economic strata from Renigunta, Andhra Pradesh considerably lower. A dietary survey by Sharma and Sharma (1977) among the Khond tribe of Orissa indicated that the intake of protein, fat, and carbohydrates falls far short of the desired levels. Hence, the basic calorie requirements are not met. Sharma (1978) in a longitudinal study of growth of twins found hereditary factor in both adolescent growth spurt and the age at spurt. Females show a greater genetic control than males, on the whole. In another study, Sharma (1983) noted that in the growth of height and weight as well as in their growth velocities, MZ twin pairs are consistently more than the DZ pairs. Kulkarni (1987) in a study of boys from three Maharashtrian groups found that the dimensions for the growth of lower body component is faster in rate than the girth dimensions of upper body component. Surprisingly, Kulkarni has not compared his findings with those of Sharma (1970) from Maharashtra though he has included his publication in the references. A cross-sectional study of growth among the school going children of 6 to 10 years of age belonging to Yadav caste of Varanasi, UP was undertaken by Sharma (2004). He has used 14 anthropometric measurements in order to see the pattern of growth with the advancement of age by using multivariate distance analysis. The combined effect of all the measurements showed a sharp pre-adolescent spurt among the girls between 9+ and 10+ years, whereas among the boys it was yet to start.

Thyroid hormone plays a very vital role in the development of the brain as well as the body. Iodine is very essential for the production of thyroid hormone. Lack of iodine results in different kinds of health problems that are called Iodine Deficiency Disorders. The most obvious sign of iodine deficiency is goiter, an enlargement of the iodine-storing thyroid gland in front of the neck, mental retardation, and deaf mutism. Iodine deficient women suffer abortions and even still birth. Their children may be born deformed, mentally deficient, or cretins. Ahmed and Srivastava (1988) found among the Gallongs, a tribe of Ragi Doke village of Arunachal Pradesh that about 15 per cent of the total population of the village suffered from goiter in varying degrees. There were cases where goiter was associated with retarded growth, feeble mindedness, and deaf mutism.

*The Hindustan Times'*correspondent reported on March 20, 1994 that Dr Basil, S. Hetzol, Executive Director of the International Council for Control of Iodine Disorders, observed that in India alone 167 million people, are at risk of IDD, about 44 million have goiter, 2.2 million suffer from cretinism, and 6.6 million suffer from neurological defects. The total number of still births and neonatal deaths are attributable to iodine deficiency. He stressed the need for total elimination before the cost becomes unmanageable. The government of India did launch a programme to provide iodized salt to identified endemic areas. In fact, not a single state in India was free from "iodine deficiency".

There are not many studies on age changes. Bagga (1998) attempted a comparative study of the pattern of age changes in the migrant Punjabi and Sindhi women, and the local Maharashtrian women. Weight increase and general obesity among the Punjabi and Sindhi women are due to the habitual excessive food intake, consumption of richer fried and frequent non-vegetarian food, compounded by sedentary lifestyle. On the other hand, lesser consumption of calcium rich food such as milk and milk products by the local Maharashtrian Brahmin women may be responsible for a rapid reduction in bone mass indicated by sharp decrease in linear measurements such as stature and sitting height. Bagga and Sakurkar (2003) also estimated the age changes in body composition of adult Maratha women from the mental hospital in Pune. They found increase in the body weight, body circumferences, and skin fold thickness with increasing age. It is not clear why the authors have chosen subjects from mental hospital.

The hypothesis pertaining to association of blood groups and disease means that if an invading organism possessed an antigen similar to one of the blood groups antigen, a carrier of the latter might have difficulty in elaborating an antibody against the invader (Mourant et al., 1976). The first convincing data came from the investigation of Aired et al. (cited in Mourant et al., 1976) who found that A was in highly significant excess among the sufferers from carcinoma of the stomach. Vogel and Chakravarti (1966) carried out investigations in Burdwan district of West Bengal where they found positive relationship between groups A and AB, so far as the susceptibility to small pox is concerned. But, they did not find any association between the ABO blood groups and the immunological type of leprosy (lepromatous vs non-lepromatous). Ramchandraiah and Shaikhavali (1978) did not find any relationship between ABO blood groups and jaundice, filariasis, and diabetes mellitus, though in the latter an excess of phenotype A was observed. There are many investigations in India where association between blood groups and pulmonary tuberculosis was assessed. But what is most important is that where the probable or possible associations with the ABO groups and diseases have been claimed, according to Chung and Morton were not conditions frequently causing death before or during the reproductive period; hence they are not likely to have selective effect (cited in Mourant et al., 1976).

There have been attempts by Indian physical anthropologists to see whether or not there exists any relationship between disease and dermatoglyhpic traits. Chakravarti while researching as a Humboldt Fellow did not find association between diabetes mellitus and dermatoglyphics. Bagga (1987) has cited many investigators, including herself who have found contradictory results as regards the association of schizophrenia and dermatoglyphics. Some have found same frequency of axial triradii in both patients and controls while others found higher incidence in the patients. It has also been found that model types of C line terminations show marked differential trends among Indian populations. The use of dermatoglyphics as a diagnostic tool has generally been accepted and its importance in the diagnosis of Down's syndrome cannot be denied. However, I am not impressed by those who just report deviations from normal frequencies while fixing the association between dermatoglyphics and congenital abnormalities. Such findings lead us nowhere as we cannot diagnose a disorder on this basis. The diagnosis is possible only where the pattern or combination of patterns is unique or exceedingly rare in the normal population. I am reminded of Goming's (1968) observation that even statistical correlation cannot be free from pitfalls. The more correlations are made, the more likely it is that some will be found to be statistically significant. If none were found, this statistical rarity is worth reporting. Hence, excessive claim by those working in this area may reduce the science of dermatoglyphics to non-science.

Mathew (1979) has found significant difference in stature among the Indian soldiers of ABO blood groups. He observed that B group individuals are tallest and those of A group shortest. Jolly et al. (1960) in a sample of 537 donors at the Blood Bank, Lucknow Medical College found no association between stature and ABO blood groups and as such their results are not in conformity with that of Mathew. However, Jolly et al. made certain observations with regard to the Mathew's and their own study: (i) in neither case the subjects were selected according to standard statistical techniques, (ii) in the case of Mathew's study, the subjects represent a sample from truncated population with regard to stature and this fact should be taken into account in the analysis.

In another investigation, Srivastava et al. (1984) did not find any significant difference in the stature and ABO blood groups of post-graduate students of Karnataka University, Dharwad, who were above the age of 20 years. In the same study, it was also found by the resident doctor of the university that rural male students showed lower incidence of dental caries than the urban ones, and the caries were mostly confined to the lower jaw in both the groups of students. Flurosis was found in five per cent of rural boys as compared to the urban boys. Tartar was present in 80 per cent of both the groups. Only three cases of pigeon chest were noticed. Yusuf et al. (1980) reported higher prevalence of dental caries among 500 Muslims of Andhra Pradesh, a neighbouring state of Karnataka. It is higher (42.4%) in the urban sample than in the rural (34.8%). They also pointed out that the frequency of caries is lower in flurosis-affected persons. Data obtained by other investigators from some castes indicate the variation of dental caries from 29 to 32 per cent, highest being 50 per cent in Kshatriyas of AP.

Malocclusion encompasses all deviations of the teeth and jaws from normal alignment. Malocclusion clusters around 34 per cent though the urban males have 17 per cent. Awaradi (1979) in his sample from Karnataka also found similar percentage (24) of Malocclusion. Brash et al. (1956) have suggested that malocclusion is inherited. Orthodontists often emphasize the environmental determinant of malocclusion such as oral habits, abnormal swallowing patterns, and the premature loss of deciduous teeth. Some anthropologists have pointed out that technologically primitive people have better aligned dentition than civilized people because people living in primitive conditions with a supposedly rougher and functionally more stimulating diet achieved this growth potential in tooth and jaw without malrelations. There are several investigations on the eruption of teeth, falling of teeth, dental morphology and pathology, and odontometric analysis. They are in scattered populations, hence nothing much can be said.

Inheritance of defective traits has been investigated by physical anthropologists. Sharma (1977) reported that brachydactyly, abnormal segmentation of the fingers and toes, ectodactly, and lobster-claw involving both hands and feet are related defects which are inherited due to an autosomal dominant mutant gene. Sharma's results are in conformity with those of Tubby (1894) and Lewis and Embelton (1908) but the mode of inheritance is not fully established as the anomalies are highly variable and overlapping. According to Chattopadhay (1980), there is no general agreement about the mode of inheritance of ear lobe type in man. It has been reported by Hilden (1922) that it is controlled by dominant gene, while Carrier (1922) suggested recessive gene. Chattopadhay concluded that ear lobe types are controlled by a pair of allelomorphic autosomal gene. The gene for free type is dominant over the gene for the attached type. Most of the authors have found the inheritance complicated.

In two significant studies, Das (1958) and Gai (1982a, 1982b) have tested the hypothesis of incomplete penetrance of the dominant allele in the inheritance of PTC and Palmaris Longus muscle, respectively. Palmaris Longus muscle is a slender fusiform muscle lying on the medial side of flexor carpi radials. The muscle is often absent and its frequency is related to a racial background and sex. Gai also noted that there is a disagreement among the investigators (Thomson et al., 1921; Dansforth, 1924; Zebrowaki, 1934; Suzuki, 1957; Kusaji, 1960) as regards the mode of inheritance. Gai has analysed his data by using the (1) Fisher's method of family analysis and (2) a prior expectation of recessives in sibship of different sizes containing atleast one recessive child. According to him, the degree of penetrance of the dominant allele controlling the Palmaris Longus muscle in the heterozygote was 60 per cent in GX G type of mating and 65 per cent in GX g type of mating. Hence, the gene responsible for Palmaris Longus muscle has high penetrance (1982b). Gai (1987) also observed the level of penetrance of the gene controlling the tongue rolling. His study revealed that the tongue rolling is inherited by a dominant gene which has incomplete penetrance. The penetrance value in the RX R type of mating was 25 per cent and 35 per cent in both RX r and rX r type of mating.

The physiological processes controlling the reproductive function differ in the two sexes of human species. For the females, it is the period between menarche and menopause that constitutes the reproductive age of their life. It has been reported by various investigators that menarche and menopause vary from population to population due to factors such as nutrition, environment, and heredity. Tanner (1960) opines that menarcheal age depends on the combined action of genes at many different loci rather than on any single gene. Identical twins experience their first menstruation within few weeks of each other. As early as 1895, Robertson investigated the period of puberty in Hindu women of India. Curjel (1920) emphasized the "knowledge of the average age of the onset of puberty, its association with early marriage and the average menstrual life is desirable for correct treatment of morbid gynecological conditions with these ages". Sen (1953) is the first Indian physical anthropologist to study the mean menarcheal age among the Bengalee women. Pillai and Srivastava (1979) compiled the data from different populations of India. Normally, the average menarcheal age centres around 12 years but there are certain groups in which the mean is higher. These are Brahmins from Maharashtra and Nayars from Kerala. The lowest mean menarcheal age (11.71 years) is observed in Assamese girls (Srivastava et al., 1968) whereas Rakshit (1960) reported 12.39 years. For the sake of convenience, Pillai et al. tabulated the available data into three regions of the Indian subcontinent. They put Uttar Pradesh, Bengal, Assam, Orissa in the northeastern region and Maharashtra, Karnataka, and Kerala in the southwestern region. It seems that the girls of northeastern region reach menarche earlier than the girls of southwestern region as is evident from the means of these two regions. The average comes to 12.88 years in the former and 14 years in the latter.

Offspring of marriages between close relatives are more likely to show traits controlled by recessive genes than the children of unrelated persons. The common usage for such marriages is consanguinity. Consanguinity reflects on the inheritance of recessive traits, increases the incidence of an X-linked trait in females, and has its implications on morbidity and early mortality. The extensive study on inbreeding is that of Sanghvi (1966) in rural areas of Andhra Pradesh. There is a great preference of consanguineous marriages in South India. Sanghvi found high proportion of uncle-neice and matrilateral cross-cousins types of marriages. In fact, there were two out of every five marriages. There were also 2.13 per cent patrilateral cross-cousin types of marriages. While discussing the effects of inbreeding in his study, Sanghvi pointed out that the frequency of deleterious alleles would have considerably declined in a population such as that of Andhra Pradesh under inbreeding over a period of two millennia or perhaps longer. Rao and Indaraj (1977) found in North Arcot district of Tamil Nadu that out of total number of marriages 47 per cent in the rural and 29 per cent in the urban areas were consanguineous. In more than 80 per cent of consanguineous marriages, the spouses were first cousins or more closely related. There is another series of data from Tamil Nadu, which Roychoudhury (1980) collected from the monographs of 1961 Census. It refers to different religious communities from 48 villages spread over 12 districts. The Muslims showed higher rate of consanguinity than the Hindus. The consanguinity among the three categories of Hindus, namely, high caste, scheduled caste, backward caste, and scheduled tribe varied from 26.2 to 31.9 per cent. There is not much difference in the frequency of uncle, niece and first-cousin marriages among these categories.

Congenital malformation according to Verma and Methews (1983) comprised 8 per cent of the prenatal mortality in India. As regards the specific malformation, it has been found that Sikhs have a high incidence of neural tube defects. Prevalence of anencephaly and spinabifid is high in Punjab, Delhi, and Rajasthan as compared to southern and eastern states of India. Actual cause of common congenital malformations is not known, perhaps both hereditary and environmental factors are involved. Sarkar (1941) has written about the Natural Eugenics, where he says nature helps in eliminating a large number of hereditary defects. Deaf-mutes are short-lived, hemophiliacs rarely live beyond the age of 20–22 years. Some of the grave anomalies, such as Xeroderma pigmentosum, Ichthyosis, congential Huntington's chorea, amaurotic family idiocy, etc. are also the causes of early extinction of the individual. The death rates of low-grade imbeciles and idiots, insanes, and epileptics are relatively high. The foregoing description shows the utility of eugenics and as such, eugenic education is primarily needed. The initial eugenic programme should begin with:

1. sex education for youth,
2. preparation for marriage, and
3. marriage counselling as envisaged by Belisle (cited in Sarkar, 1941).

Sarkar (1943–44) also collected data from various Indian hospitals to ascertain the frequency of twin and other multiple births, triplets and quadruplets. Statistics from 30 hospitals outside Bengal shows 2,582 twins in a total number of 209,023 pregnancies giving a ratio of 1:80.8, whereas 11 Bengal hospitals yielded a ratio of 1:75.3 obtained from 1,385 twins of 104,290 total births. Ceylon presented a ratio of 1:161.1 derived from 10,056 twins in 1,620,077 total births. The ratio for India was found to be 1: 79.0 or 1.27 per cent. Thirty-seven triplets from hospitals outside Bengal, nine from Bengal, and eight from Ceylon (now Sri Lanka) were reported. Triplets show an excess of females (66:39) over males. On the basis of Weinberg's rule, it was found that 64.7 per cent and 35.3 per cent of the total number of twins were dizygotic and monozygotic, respectively. In another study, Venkateshwara Rao (1978) collected data from two hospitals in Vishakapatnam town. Out of 10,916 deliveries, 146 were multiple births, comprising 144 twins and 2 triplets. The overall rate of multiple births comes to 1.32 per cent. Significant effects of maternal age and parity on twinning were also observed. Younger mothers exhibited low rates of twinning in low parities and higher rates in high parities. Older mothers showed high rates of twinning right from the low parities with a declining rate after parity 4.

There are certain areas in which more attention to population analysis appears promising. One such area is chromosome variation where human chromosomes stain differently and show characteristic banding patterns (Roberts, 1980). In India, Ghosh and his colleagues are doing research in this area. The reduced frequency of sister exchanges in human lymphocytes cultured with autologous serum and lateral asymmetry of constitutive heterocharomatin in human chromosomes may be cited as some of their investigations (Ghosh and Nand, 1979).

The skin colour in racial crosses was studied by subjective methods of visual scoring or matching against some artificial standard. Tiwari (1963) used reflecometry technique in the quantitative assessment of skin pigmentation in 23 families in which the father was Indian and the mother European. It was noted that the mean values for the hybrids deviate only significantly from the mid-parent values. He found similar reflectance values in the Bengalees and the miscellaneous Indians. Brahmins appeared slightly lighter.

Bird in as early as 1953–54 pointed out that reaction of seed extracts with red cells can be useful in distinguishing blood samples from several animal species. Hence, the possibility of using lectin in forensic investigation is there (cited in Bhalla et al., 1978). Bhalla et al. (1979) have used extracts of 80 Indian seed varieties belonging to 45 plant families in testing precipitin activities against the sera of 12 vertebrate species. In another investigation of this nature, they found that some lectins are useful in distinguishing animal species from one another on the basis of reaction with red cells.

Application of Physical Anthropology in readymade-garment industries, architecture and town planning, people's health, comfort, and safety have been pleaded by Verma (1958a, 1958b, 1969). He anticipated the readymade garment industry would develop in India. An increasing number of people are expected to cultivate the habit of purchasing made-to-wear dresses. Therefore, it is necessary for manufacturers of such garments to standardize the size and assure the best possible fit for maximum number of people from different regions of the country. He has suggested that trained anthropmetricians take measurements and statisticians must develop special techniques that can be used by clothing technologist. In another article, Verma (1969) emphasized that physical anthropologists should be interested in determining the space required by man for different purposes and the utilization of space to the utmost advantage. Useful advice in matters of housing design can come not only from the architects, engineers, and planners but also from anthropologists. Srivastava (1963b) pleaded for applied physical anthropological research in India where he cited Le Gros Clarks' (1951) observations "A new branch of physical anthropology coming to be known as ergonomics being developed in the Department of Anthropology at Oxford. Ergonomics according to him stresses the need for fitting the machine to the man so that the former could be humanized rather than the latter in mechanized". Malik and Gakhar's studies (1998) reveal that VTD (video terminal display) users suffer from work related stresses. Between 17 and 22 per cent respondents suffer from headaches, neck pain, upper-back pain, and lower-back pain almost always. Less than 10 per cent claim blurred vision. The height of chairs and tables are not optimal for prolonged VTD use which maybe the cause for various types of pains.

There are many hominoid fossil remains from Siwalik hills and Narmada valley. The mandible discovered by Simons and Chopra (1969) has been named by them as a new species *Gigantopethicus Bilaspurensis*. A skull cap found by Sonakia (1984) at Hoshangabad, Narmada valley has been studied by many experts.

On the basis of morophometric and comparative studies, Kennedy et al. (1991) earlier and later also opined that the Narmada skull cap was that of *archaic Homo sapiens* with some features of Asian *Homo erectus*. There are varying views pertaining to Narmada fossil, as some authors believe it to be of *Homo erectus*. Sankhyan (2007) suggested that there are two possibilities, either we assign it to *archaic Homo sapiens* on grade basis, or to a different species on clad basis. His emphasis was on the latter for a new nomenclature, *H. narmadensis*. In view of the controversy of Narmada Calvaria, creation of a new species needs more convincing evidence.

It has not been possible to mention every author in the text. I assure them that their findings were not ignored. Those social anthropologists who promote the idea of taking away the bones and stones from anthropology forget that the strength of anthropology lies in its integrative nature. In this context, I wish to quote few well-reputed scholars of anthropology. Hulse (1971) has beautifully said that "the cultural nature of man is an aspect and product of his biological nature and his biological nature is an aspect and product of his cultural nature". Professor M.N. Srinivas while inaugurating the Silver Jubilee Celebration of the Southern Regional Office at Mysore, decried the distinction being made between social anthropologist and sociologist (1986). While emphasizing the unity of the Science of Man, Clyde Kluckhohn observed that "every social anthropologist must familiarize himself with genetics and other aspects of human biology if he wishes to be in the strict sense an Anthropologist" (cited in Langhlin, 1963). Sol Tax (1975) opined, "since our species' biology is intimately related to the historical, social, and cultural differences which affect individual population, human biology is a unique part of general anthropology".

McCown emphasized "it seems to me that criticism along this line coming with least grace from cultural anthropologists because demonstration that man is a mammal, that he is a primate, and hence add lustre and uniqueness to their enquiries into a man as a maker and practitioner of culture" (cited in Kennedy, 1984).

References

Adam, A., Doron, D., and Modan, R. 1967. Frequency of Protan nad Deuteran alleles in some Israeli communities and a note on selection. *Am. J. Phys. Anthrop.* 26: 297–305.

Aird, I. 1953. The association between the blood group and gastro-intestinal disease. *Gastroenterolia (Basal)* 92: 95–99.

Ansari, G. 1959–60. Muslim caste in Uttar Pradesh: a study of culture contact. *The East. Anthrop.* Spl. No. 13: 1–85.

Awaradi, S.A. 1979. Pattern of tooth occlusion. *Ind. J. Phys. Anthrop. Hum. Genet.* 5: 85–88.

Bagga, A. 1987a. Palmar Pattern and axial triradii among the schizopirenias. *Man. In. Ind.* 67: 251–259.

Bagga, A. 1987b. Courses of C line in schizophrenia. *Int. J. Anthrop.* 2: 157–166.

Bagga, A. 1998. Normality of aging: a cross cultural perspective. *J. Hum. Ecol.* 9: 35–46.

Bagga, A. and Sakurkar, A. 2003. Estimation of age changes in body composition of adult Indian women. *Anthrop. Kozl.* 44: 37–47.

Balakrishnan, V. 1978. A preliminary study of genetic distance among some populations of the Indian subcontinent. *J. Hum. Evol.* 7: 67–75.

Balakrishnan, V. 1981. Genetic distance studies in India. *Acta. Anthropogenatica.* 3: 59–65.

Basu, A. 1969. The Pahira: a population genetical study. *Am. J. Phys. Anthropo.* 31: 399–416.

Begg, P.R.1954. Stone Age man's definition. *Am. J. Orthodontist.* 40: 298–312.

Bennet, K.A. 1969. The typological versus the evolutionary approach in skeletal population studies. *Am. J. Phys. Anthrop.* 30: 407–414.

Bhalla, V. 1966. Blood group distribution pertaining to ABO, MNSs and Rh-hr system in the Indian sub-continent. *Anthropologies* 5: 67–80.

Bhalla, V., Arora, V.K., Jain, N., and Gaur, V. 1979. Lectin studies VI. selective precipitation of sorum proteins by extracts in agar gel. *J. Ind. Acad. Foren. Sci.* 18: 50–56.

Bhalla, V., Gaur, V., and Bhatia, K. 1978. Lectin studies III. A survey of phytohemagglutinins of lectins with erythroyes of ten vertebrate species. *Vox. Sanguins*. 35: 241–247.

Biswas, P.C. 1936. Uber hand and Finger – Leisten vonlundern. *Z. Morphol. Anthrop.* 35: 519–550.

Biswas, P.C. and Bhattacharya, D.K. 1966. Growth trends in Northern Indian population: variation of four Anthropometric traints from birth to maturity (cross-sectional). *Proc. Hum. Adap. Environ. Phys. Fitness.* Ed. M.S. Malhotra. 270–280.

Biswas, P.C. and Chaudhari, B. 1938. On the Palmar prints of the Oriyas. *J. Dept. Sci. Uni. Calcutta* 1: 111–129.

Brash, J.C., Mc Keag. H.T.A., and Scot, J.H. 1956. *The Aetiology of Irregularity and Malocculusion of Teeth.* UK: The Dental Board of UK.

Busi, B.R., Veerraju, P., and Rao, V. 1979. Studies on some red cell enzyme group systems among Andhras. *Acta. Anthropogenetica.* 3: 211–218.

Chattopadhyay, P.K. 1980. Inheritance of the ear lobe types in man. *Ind. J. Phys. Anthrop. Hum. Genet.* 6: 13–17.

Chattopadhyay, P., Dutta, P., and Kahsyap, V.K. 2001. Population data for nine florescent based STR loci among four important tribal populations in India. *J. Forens. Sci.* 96: 184–188.

Chattopadhyay, P. and Kashyap, V.K. 2001. Distribution of alleles of three tetrameric STR loci in eight predominant populations of India. *Foren. Sci. Internat.* 119: 134–137.

Chattopadhyay, P., Kashyap, V., and Vasula, T.S. 2001a. Genomic variation at three tetrameric STR loci among eight ethnic population groups of India. *Hum. Biol.* 71: 439–446

Chattopadhyay, P., Kashyap, V.K., and Vasula, T.S. 2001b. Gene diversity at three tetrameric STR loci among eight ethnic populations of West Bengal, and Manipur, India. *Int. Hum. Gene.* 1: 255–262.

Clark, Le Gros. 1958. Reorientation in physical anthropology. In D.F. Rober, J.S. Weiner (ed.) *The Scope of Physical Anthropology and its Place in Academic Studies.* New York: The Wenner Gren Foundation for Anthropological Research.

Coon, C.S. Garu, S.M. Birdsell, T.B. 1950. *Racos.* Springfield, Ill. Thomas.

Coon, Carlton S. 1969. *The Living Races*. New York/Canada: Alfred A Knoff Inc.

Curjel, D.F. 1920. The reproductive life of Indian Women. *Ind. J. Med. Res.* 8: 366–371.

Darrow, 1938. and Levine and Stetson. 1939. In Carter, C.O. (1962) *Human, Heredity*, Penguin Books Ltd. Middlesex.

Das, B.M. 1979. Physical variation in three Assamese castes. *Anthrop. Anz.* 3: 201–210.

Das. B.M., Das, P.B., Walter, H., and Dander-Hopfe, R. 1985. Anthropological studies in Assam, India. Observations on five Mongoloid populations. *Anthrop. Anz.* 43: 193–204.

Das, B.M., Deka, R., and Das, R. 1980. Haemoglobin E in six populations of Assam. *Ind. J. Anthrop. Soc.* 15: 153–156.

Das, B.M. and Deka, R. 1985. Population study in Assam; ABO blood groups, Haemoglobin E and G6PD deficiency. *Anthrop. Anz.* 43: 81–86.

Das, B.M.,.Walter, H., Gilbert, K., Linberg, P., Malhotra, K.C., Mukherjee, B.N., Deka, R., and Chakrabarthy, R. 1987. Genetic variation of five blood group polymorphisms in the populations of Assam, India. *Int. J. Anthrop.* 2: 325–340.

Das, S.R. 1958. Inheritance of PTC taste character in man. An analysis of 129 Rahri Brahmin in West Bengal. *Ann.Hum.Genet* 20: 200–212.

Dobzhansky, T. and Dunn, L.C. 1958. *Principles of Genetics*. 5th Edn. Mcgraw-Hill Book Co., New York.

Dutta, R. and Kashyap, V.K. 2001. Allele frequency of six STR loci among three predominant populations of Eastern India. *Foren. Sci. Int.*119: 129–130.

Eveleth, P.B. and Tanner, J.M. 1976. *Worldwide Variation in Human Growth*. Cambridge: IBPS Cambridge University Press.

Fisher, R.A. 1930. *Genetical Theory of Natural Selection*. Oxford: The Clarendon Press.

Gai, P.B. 1982a. Inheritance of Palmaris longus musele in Man. *Man in India* 62: 271–294

Gai, P.B. 1982b. Pamaris longus muscle with incomplete penetrance dominant gene. *Man in India* 62: 271–294.

Gai, P.B. 1987. Incomplete penetrance of tongue rolling. *Ind. J. Phys. Anthrop. Hum. Genet.* 13: 87–101.

Ghosh, P.K. and Nand, R. 1979. Reduced frequency of sister chromatid exchanges in human lymphocytes cultured with autologous serum. *Hum. Genet.* 51: 167–170.

Gomings, D.E. 1968. Cancer and Dermatoglyphics. *Lancet* 1: 1428.

Guha, B.S. 1931. Racial affinities of the people of India. Census of India, Vol. I, Part III Ethnological.

Gulati. R. 1971. Inter-group differences in an Artisan caste of Maharashtra. In Harry Field (ed.) *Coconut Groove*. Miami, FL: Field Research Projects..

Haldane, J.B.S. 1932. *The Causes of Evolution*. New York: Harpers.

Hashin Ahmed, S. and Srivastava, R.P. 1988. Iodine deficiency: a tale of misery of Gallongs of village Ragi Doke, Siang division, Arunachal Pradesh. *J. Hum. Ecol.* 9: 93–76.

Heidi, D., Das, B.M., Walter, H., Das, P., and Das, R. 1988. Anthropological studies in Assam, India: differentiation processes among Assamese populations. *Anthrop. Anz.* 46: 159–184.

Hulse, F.S. 1971. *The Human Species. An Introduction to Physical anthropology*. New York: Random House.

H.T. Correspondent. 1994. Big Challenge of Iodine Deficiency. *The Hindustan Times,* March 20.

Jolly, J.G., Manglik, N.S., and Srivastava, R.P. 1960. Blood groups and stature. *Ind. J. Path. Bact.* 3: 180–185.

Karve, I. and Dandekar, V.M. 1951. *Anthropometric Measurements of Maharashtra,* Deccan College Monograph Series. 8.

Karve, I. and Malhotra, K.C. 1968. A biological comparison of eight endogamous groups of the same rank. *Curr. Anthrop.* 9: 109–116.

Katakkar, S.M. 1992, *Encyclopedia of Palm, and Palm Reading*. New Delhi: UBS.

Kaur, H. and Chattopadhyay, P.K. 1998. Haptoglobin types from blood stains. *J. Ind. Anthrop. Soc.* 35: 11–14.

Kennedy, A.R. 1984. A reassessment of the theories of racial origins of the people of the Indus Valley civilization from recent anthropological data. In K.A. Kennedy and G.L. Posehl (ed.) *Studies in the Archaeology and Paleoanthropology of South Asia*. New Delhi: Oxford and IBH.

Kennedy, K.A.R., Brooks, S.T., and Mc. Cown, T. 1984. A perspective on a physical anthropologist. *The Wenner Gren Foundation for Anthropological Research* 15: 103.

Kennedy, K.A.R., Sonakia, A., Chiment, J., and Verma, K.K. 1991. Is the Narmade hominid an Indian *Homo erectus*. *Am. J. Phys. Anthrop.* 86: 175–196.

Kirk, R.L., Lai, L.Y.C., Vos, G.H., Wiekremasinghe, R.L., and Perera, D.J.B. 1962. The blood and serum groups of selected populations in South India and Ceylon. *Am. J. Phys. Anthrop.* 20: 487–497.

Kulkarni, V.S. 1987. Assessment of mid growth spurt. *Man in India* 67: 196–219.

Kumar, N. 1965. ABO blood groups and sickle cell trait investigation in Madhya Pradesh Ratlam and the adjacent districts (Central India). *Bull. ASI* 14: 40–44.

Kurulkar, G.M. 1958. *Variation in Man. Address of the President. Anthropology and Archaeology*. Madras: Indian Science Congress.

Langhlin, W.S. 1963. Teaching of physical anthropology: concepts and *cheological Section* problems. In G. Mandel Baum et al. (eds) *The Teaching of Anthropology*. Berkeley/Los Angeles: University of California Press.

Livingstone, F.B. 1980. Natural selection and random variation in current developments. In James H. Mielke and M.H. Crawford (ed.) *Anthropological Genetics: Theory and Methods*. New York: Plenum Press.

Livingstone, F.B. 1985. *Frequencies of Hemoglobin Variants Thalassemia, the Glucose-6-Phospate Dehydrogenase Deficiency, G6PD Variants and Ovalocytosis in Human Population*. Oxford: Oxford University Press

Mahalanobis, P.C., Majumdar, D.N., and Rao, C.R. 1949. Anthropometric Survey of the United Provinces, 1941 Statistical Study. *Sankhya* 9: 9–324.

Majumdar, D.N. 1950. *Race Realities in Cultural Gujarat*. Bombay: Gujarat Research Society.

Majumdar, D.N. 1956. *Rupkund: Mystery Lake in Himalayas* (Unpublished).

Majumdar, D.N. 1961. *Races and Cultures of Cultures of India*. 4th Ed. Bombay: Asia Publishing House.

Majumdar, D.N. and Bahadur, S. 1952. ABO blood groups in India. *Eastern Anthrop.* 5: 101–122.

Majumdar, D.N. and Kishan, K. 1947. Blood group distribution in United Provinces. *The East. Anthrop.* 1: 1: 8.

Majumdar, D.N. and Rao, C.R. 1960. *Race Elements in Bengal. A Quantitative Study*. Calcutta: Asia Publishing House.

Malik, S.L. and Gakhar, I. 1998. Man-machine relationship: the human factor in information technology. *J. Ind. Anthrop. Soc.* 33: 229–241.

Malleshwara Rao, P., Balke, N.M., and Veerraju, P. 1978. Genetic studies on Savera and Jatapu Tribes of Andhra Pradesh. *Hum. Hered.* 25: 122–131.

Mathew, T.N. 1959. Relation between stature and blood groups among Indian Soldiers. *Sankhya*. 21: 1–17.

Mitra, S. 1949. *The Vision of India*. Calcutta: Jaico Publishing House.

Mourant, A.E. 1954. *The Distribution of the Human Blood Groups*. Oxford: Blackwell Scientific.

Mourant, A.E., Kopec, A.C., and Domaniewska-Sobzek, K. 1976. *The Distribution of Blood Group and Other Polymorphisms*. Oxford: Oxford University Press.

Mukherjee, B.N., Walter, H., Malhotra, K.C., Chakraborty, R., Banerjee, P.S., and Roy, M. 1985. Population genetic study in ten endogamous groups of West Bengal. *India Anthrop. Genet*. 45: 239–254.

Mukherjee, N., Mitra, M., Charkaborty, M., and Majumdar, P.P. 2000. Congruence of genomic and ethnolinguistic affinities among four tribal populations of Madhya Pradesh (India). *J. Genet.* 79: 41–46.

Mukherjee, N., Majumdar, P.P., Roy, M., Dev, B., Chakraborty, M., and Banerjee, S. 1999. Genomic variation at four short tandom repeat loci in eight population groups of India. *Hum. Biol.* 71: 439–446.

Munshi, K.M. 1974. *Foundations of Indian Culture*. Bombay: Bharatiya Vidya Bhavan.

Negi, R.S. 1963. ABO blood groups, Sickle cell trait, and colour blindness in the Maria of Baster, Gond and Kanwar of Bilaspur. *Bull. ASI* 12: 149–153.

Negi, R.S. and Das, A. 1962. The blood groups (ABO, MN and Rh) ABH secretion in Saliva and colour blindness in the Rajput of Western U.P. and Dholpur. *Bull. ASI* 11: 221–231.

Negi, R.S., Srivastava, A.C., and Bhatnagar, B.R. 1972. Distribution of ABO groups in central and western Himalayan population. *Bull. ASI* 21: 57–76.

Pillai, S.S. and Srivastava, R.P. 1979. Age at Menarche in populations of Karnataka. *Ind. Anthrop.* 9: 59–64.

Post, R.H. 1962. Population differences in red green colour vision deficiency. *Eug. Quart.* 9: 131–146.

Rakshit (Guha) Sipra. 1960. A Short note on the menarcheal age of Assamese girls. *Man India* 40: 50—55.

Ramachandraiah, T.R.K.V. and Shaikhavali, B. 1978. ABO blood groups in relation to Jaundice, Filariasis and Diabetes Mellitus. *Med. Genet. India* 2: 241–247.

Rao, S.S. and Indaraj, S.G. 1977. Inbreeding efforts on human reproduction in Tamil Nadu of South India. *Ann. Hum. Genet.* 41: 87–98.

Ridley, G.N. 1947. *Man Studies Life: The Story of Biology*. London: Watts.

Risley, H. 1915. *The People of India,* 2nd ed. Calcutta and Simla: Thacker Spink.

Robertson, J. 1895. On the period of puberty in Hindu women (cited in Curjel, D.F., 1920).

Roberts, D.F. 1970. *Climate and Human Adaptability*. Cumings Publications America.

Roychoudhary, A.K. 1980. Consanguineous marriages in Tamil Nadu. *J. Ind. Anthrop. Soc.* 15: 167–174.

Roychoudhary, A.K.1983. Genetic polymorphisms in human populations in India. In *Some Genetical Aspects*. New Delhi: ICM.

Sanghavi, L.D. 1953. Comparison of genetical and morphological methods for a study of biological differences. *Am. J. Phys. Anthrop.* 11: 385R-404.

Sanghavi, L.D. 1966. Inbreeding in rural areas of Andhra Pradesh. *Ind. J. Genet.* A Symposium No. 26: 351–365.

Sankhyan, A.R. 2007. Hominid fossil finds in Indian: current status In A.R. Walumbe, P.P. Joglekar, and K.K. Basa (eds) *Anthropology for Archaeology*. Pune: Deccan College, Post-Graduate Institute, Publications.

Sarkar, S.S. 1941. Aims and objects of eugenic researches in Bengal. *Ind. Eug. Soc. Bull.* 1: 2–19.

Sarkar, S.S. 1943–44. The frequency of plural and multiple births in India. *Trans Bose. Res. Inst.* Calcutta 16: 2–9.

Sarkar, S.S. 1954. *The Aboriginal Races of India*. Calcutta: Bookland Ltd.

Sarkar, S.S. 1962. Dermatoglyphics in India. In T.N. Madan and Gopala Sarana (eds) *Indian Anthropology*. Bombay: Asia Publishing House.

Sen, D.K. 1960. Blood groups and haemoglobin variants in some upper castes of Bengal. *J. Royal. Anthrop. Inst.* 10: 43–45.

Sen, D.K. 1962. The racial composition of Bengalis. In T.N. Madan and Gopala Sarana (eds) *Indian Anthropology*. pp. 93–220. Bombay: Asia Publishing House.

Sen, D.K. 1964–65. Races of Ancient India, A Study of Methods. Presidential Address, Section of Anthropology and Archeology. *Ind. Sci. Cong.* 51–52 session, Part III, I–23.

Sen, D.K. and Pal. 1968. Effect of genetic drift and gene flow in several endogamous groups of a village in West Bengal. *Bull. ASI* 17: 209–21.

Sen, T. 1953. Reproductive life of some Indian women. *Man India* 33: 31–54.

Sharma, A. 1983. Haemglobinopathies in India. In *Peoples of India Some Genetical Aspects*. pp. 31–49. New Delhi: ICMR.

Sharma, J.C. 1963. Age changes in bodily proportions of Maharashtrians, unpublished Ph.D. Thesis, Punjab University.

Sharma, J.C. 1977. Inheritance of defective extremities. *Ind. Med. Res.* 66: 120–127.

Sharma, J.C. 1978. Genetic contribution of adolescent growth spurt in nine anthropometric characters. A study based on longitudinal growth data on twins. Prog. *Clin. Boil. Res.* 24: 223–233.

Sharma, J.C. and Sharma, K. 1977. Nutritional anthropology of tribal community of Orissa with emphasis on morbidity and nutrition. *Kerala Sociol.* 5: 43–51.

Sharma, M.B. 2004. Multivariate distance analysis for growth characters during pre-adolescence among Yadav Children of Eastern Uttar Pradesh. *J. Ind. Anthrop. Soc.* 30: 67–76.

Shukla, B.R.K. and Srivastava, R.P. 1969. A quantitative study of finger dermatoglyphics of the Brahmins of Uttar Pradesh. *J. Ind. Anthrop. Soc.* 4: 165–173.

Simons, E.L. and Chopra, S.R.K. 1969. Gigantopethicus (Pongidae Hominodae) a new species from North India. *Postilla* 138: 1–18.

Singh, I.P. and Bhasin. M.K. 1979. Ethnic and geographic frequency distribution of dermatoglyphic traits in India. A review of birth defects. Original Article Series. 15: 347–387. *The National Foundation.*

Singh, R.D. 1961. Digital pattern frequency and size variation in some castes of Uttar Pradesh. *East. Anthrop.* 14: 169–181.

Singh, R. 1970. Growth in arm girth and other anthropometric parameters of the Punjabi boys age 11–18 years. *Z. Morp. Anthrop.* 62: 160–171.

Singh, S.K. 1994. *All Communities: The People of India* (ASI). New Delhi: Oxford University Press.

Srinivas, M.N. 1986. Special Cadre for Social Science urged. *The Deccan Herald,* May: 16.

Srivastava, A.C. and Tyagi, D.A. 1967. Study of taste sensitivity to phenylthiourea (PTC) among various population groups in Uttar Pradesh. *Bull. ASI* 16: 99–109.

Srivastava, R.P. 1959. Measurement of taste sensitivity to Phenylthiourea in Uttar Pradesh. *East Anthrop.* 12: 267–272.

Srivastava, R.P. 1961a. Frequency of non-taster among Danguria Tharus of Uttar Pradesh. *East Anthrop.* 14: 258–259.

Srivastava, R.P. 1961b. Ability to smell solutions of sodium cyanide. *East Anthrop.* 14: 189–190.

Srivastava, R.P. 1963a. A Study of fingerprints of the Danguria Tharus of Uttar Pradesh (India). *Am. J. Phy. Anthrop.* 21: 69–76.

Srivastava, R.P. 1963b. A plea for physical anthropological research in India. *J. Soc. Res.* 6: 79–82.

Srivastava, R.P. 1965a. A Quantitative analysis of fingerprints of the Tharus of Uttar Pradesh. *Am. J. Phys. Anthrop.* 23: 99–106.

Srivastava, R.P. 1965b. Blood groups in the Tharus of Uttar Pradesh and their bearing on ethnic and genetic relations. *Hum. Biol.* 37: 1–12.

Srivastava, R.P. 1966. A review of dermatoglyphic research in Uttar Pradesh. Proc. Vol. *Int. Symp. Dermat.* (New Delhi) ed. P.C. Biswas and Sharma A. 14–17.

Srivastava, R.P. and Goswami, M. 1968. Menarcheal age of Assamese girls. *The Anthropologist.* 15: 57–60.

Srivastava, R.P. 1978. Distribution of blood groups in Uttar Pradesh *Acta. Anthropogenetica.* 2: 49–58.

Srivastava, R.P. 1983. Incidence of colour blindness in Assam and its variability among the Indian populations. *Ind. Anthrop.* 13: 51–62

Srivastava, R.P. 1989–90. Modernization of teaching and research in anthropology. *Bull. Dept. Anthropol.* Dibrugarh University 18–19: 119–125.

Srivastava, R.P., Kundur, V.G., and Gai, P.B. 1984–85. A biomedical survey among the post-graduate students of Karnatak University. *J. Karnataka University* 20–21: 134–141.

Srivastava, R.P. and Preetha, K. 1979. Further data on Menarcheal age in Karnataka. *Man Life* 5: 171–174.

Srivastava, R.P., Sarana, G., Gai, P.B., Hittalamani, S.V., Awaradi, S.A., and Umapatty, B.E. 1979. Distribution of ABO blood groups in Karnataka (India). *Curr. Anthrop.* 20: 65.

Sol Tax. 1975. General Editors Preface. In E.S. Watts et al. (eds) *Bio-Social Interrelations in Population Adaptation.* The Hague: Mouton Press.

Sonakia, A. 1984. The Skull of early man and associated mammalian fauna from Narmada valley, Alluvium, Hoshangabad area M.P. (India). *Rec. Geol. Surv. India.*

Sonia, C. 1965. *Races of Man.* Printed by the order of Trustees of British Museum.

Sudhakar, Babu M., Veerraju, P., and Naidu, T.M. 1980. A note on the sickle cell trait in a tribal population of Coastal Andhra Pradesh. *Ind. Anthrop.* 10: 125–129.

Tiwari, S.C. 1960. Haptoglobin and transferring variants in some upper castes of Bengal. *Anthropologist* 7: 1–17.

Tiwari, S.C. 1961. The frequency of serum haptoglobin and transferring types in Punjabi population. *Anthropologist* 18: 23–28.

Tiwari, S.C. 1963. Studies of crossing between India and Europeans. *Ann. Hum. Genet.* 26: 219–227.

Transactions of the Bose Research Institute Calcutta. Biological and Physical Researches. xvi: 2–9.

Veerraju, P., Busi, B.R., and Anand, A. 1978. Red cell glucose-6-phosphate dehydrogenase deficiency among Andhras. *Am. J. Phy. Anthrop.* 49: 143–144.

Venkateshwara Rao, T. 1978. Maternal age, parity and twin pregnancies. *Biol. Epidem.* 99–103.

Verma, B.K. 1958a. Ready-made garments industry in India. *The Sunday Standard,* Bombay, September 7.

Verma, B.K. 1958b. Physical anthropology for peoples' health, comfort, and safety. *The National Herald,* Lucknow.

Verma, B.K. 1969. Anthropometry applied to architecture and town planning. *J. Soc. Res.* 12: 78–92.

Verma, I.C. and Methews, A.R. 1983. Congenital malformations in India. In *People of India: Some genetical aspects.* pp. 70–84. New Delhi: ICMR.

Vogel, F. and Charkaravartti, M.R. 1966a. ABO blood groups and small pox in a rural population of West Bengal and Bihar. *Hum. Genetik.* 3: 166–180.

Vogel, F. and Charkaravartti, M.R. 1966b. ABO blood groups and the type of leprosy in Indian population. *Hum. Genet.* 3: 186–188.

Walter, H. 1986. Genetic differentiation processes among the populations in India. *Int. J. Anthrop.* 1: 297–306.

Walter, H., Pal, K.P., Helling, M., Veerraju, P., Goud, J., Naidu, J.M., Sudhakar, Babu, M., and Jaikishen, G. 1981. Genetic markers in eight endogamous population groups from Andhra Pradesh (South India). *J. Morph. Anthrop.* 72: 325–328.

Walter, H., Veerraju, P., and Helling, M. 1978. Serum protein polymorphisms in three tribal populations from Andhra Pradesh. *Acta. Anthropogenetica.* 2: 25–32.

Watkins, W.S., et al. 1999. Multiple origins of the mtDNA 9–bp deletion in populations of South India. *Am. J. Phys. Anthrop.*109: 147–158.

Watkins, W.S. Prasad, P.V.R. Naidu, J.M. Rao, B.B. Bhanu, A.B. Ramachandram, B. Das, P.K. Gai, P.B. Reddy, P.C. Reddy, P.G. Sethuraman, M. Bamshad, M.J. Jorde, L.B. 2005. Diversity and divergence among the tribal populations of India. *Ann. Hum. Genet.* 69: 680–693.

Wells, S. 2003. *The Journey of Man: A Genetic Odyssey*. New York: Random House.

Wright, S. 1931. Evolution in Mendelian population. *Genetics* 16: 97–159.

Wood Jones, F. 1926. *Arborial Man.* New York: Hafner.

Yusuf, B., Reddy, K.N., and Reddy, R.V. 1980. Dental caries and flurosis among Muslims of Kurnool, Andhra Pradesh. *J. Hum. Ecol.* 9: 77–81.

5 Public Health, Human Genomics and Disease

Indera P. Singh

Public health has been defined as the health outcome of a group of individuals, including the distribution of such outcomes within the community. It is an approach to health that aims to improve the health of an entire population. One major step in achieving this aim is to reduce health inequities among population groups. Population health seeks to step beyond the individual-level focus of mainstream medicine and public health by addressing a broad range of factors that impact health on a population level such as environment, social structure, resource distribution, etc.

An important theme in population health is the importance of social determinants of health and relatively minor impact that medicine and health can have on improving health overall. There is a very robust correlation between socio-economic status and health. Lower socio-economic status has been linked to chronic stress, heart diseases, ulcers, Type 2 diabetes, rheumatoid arthritis, certain types of cancers, and premature ageing. Disease is not confined only to lower socio-economic groups but there is a continuous gradient from top to bottom of the socio-economic ladder.

India has currently the world's second largest population along with a fast-growing economy and significant economic disparity. Major population movements, social structure, and caste endogamy have influenced the genetic structure of Indian populations. The understanding of these influences is important as gene mapping and case-control studies are initiated in the Indian population. India with more than 1,080 million people has the largest population in the world and it is one of the world's top 12 mega diversity countries and has a vast diversity of human beings, fauna, flora, and environmental regimes.

Its present population includes stone-age food-gatherers, hunters, fisher-folk, shifting cultivators, peasant communities, subsistence agriculturists, nomadic herders, entertainers, as well as those engaged in mechanized agriculture, mechanized fishing, tapping offshore oil and natural gas, running atomic power plants, and producing computer software. India has been peopled by human groups carrying a diversity of genes and cultural traits. We have almost all the primary ethnic strains, i.e., Proto-Australoid, Mediterranean, Mongoloid, Negrito, and a number of composite strains. It is the homeland of over 4,000 Mendelian populations, of which 3,700 endogamous groups are structured in the Hindu caste system as *jatis*. Outside the purview of 'caste system' there are a thousand odd Mendelian populations which are tribal autochthones and religious communities.

India also has a high rate of infectious disease and increasingly higher rates of chronic disease. India cannot afford to adopt existing expensive western healthcare models to address these local health needs. India needs to look to innovative, cost-effective models including adoptive emerging health technologies. Political will and institutional leadership are the most important ingredients to achieve this aim. Political leaders of India made a beginning in 2001 when the Ministry of Health pledged 20 million dollars for research in medical genomics. Recently, the National Biotechnology Development Strategy has identified genomic medicine as an inevitable opportunity, but much more needs to be done.

Private sector especially the pharmaceutical industry also has big stakes in the development of cost-effective drugs in developing countries like India. This will also involve international cooperation. Genomic sciences can add value to local health care systems by emphasizing prediction, prevention, and possibly decreasing the cost of healthcare through better diagnosis, early detection, and improved treatment and management.

India could advance its health product innovation system and building capacity in science and technology by investing in these emerging fields of genomic health. This will provide India with a competitive edge in the knowledge-based economy, resulting in economic stimulation. The Indian Genome Variation (IGV) Consortium, a government-funded collaborative network, is a reflection of these efforts. The IGV has recently developed the first large scale database of genomic diversity in the Indian population that will facilitate the research on disease predisposition, adverse drug reactions, and population migration.

Therefore, genomic medicine is the need of the hour and India has taken the first step to develop these technologies to improve public health and create knowledge-based economies. Our revered first Prime Minister of India, Jawaharlal Nehru urged the Indian people to develop scientific temper and be independent, which can be noticed in Indian scientists dominating researches in space and information technology: "I urge the young people to continue in this spirit and be the leaders in developing innovative and alternative gene technologies".

Part II

Social Anthropology

6 The Family: A Human Imperative

Need to Save its Defining Features and Safeguard its Future

N. Subba Reddy

Everywhere in the world today there is a wide-ranging debate over the future of the family. The debate is precipitated by the far-reaching transformations that have occurred in the domestic arena. These transformations are so radical that they make one wonder whether some of the new forms share anything with the original institution. A preliminary question that arises is whether the new versions can be technically called families? In several publications in the recent years, one finds that all kinds of domestic units are given the tag, family. You come across terms such woman-headed family, grandmother-headed family, single-parent family, lesbian family, gay family, etc. There seems to be a strong need to recapitulate the definition of the word 'family' and apply it only to domestic units where the defined criteria are present and designate others by generic terms such as household, homestead, domestic unit, etc.

This brings us face to face with the question whether there is a generally agreed definition of the family and whether there have been no dissenting voices against it? So we have to consider both sides of the issue before we arrive at a conclusion. So, I shall begin with what I consider to be a generally acceptable definition of the family. Murdock (1960) is well known for his view extolling the universal importance of the nuclear family. The first two lines in his book, *Social Structure*, read as follows: "the family is a social group characterized by common residence, economic cooperation and reproduction. It includes adults of both sexes, at least two of whom maintain a socially approved sexual relationship, and one or more children, own or adopted, of the sexually cohabiting adults" (ibid.: 1). From this passage, one can identify three component dyads of the family, viz., the conjugal dyad (husband and wife), the maternal dyad (mother and children), and the paternal dyad (father and children). If we follow Murdock's definition, the combined presence of the three dyads is necessary for a unit to be termed as family. The four important functions of the family listed by Murdock are: (1) sexual, (2) economic, (3) reproductive, and (4) educational (socialization). Explaining the importance of these functions, he says: "without provision for the first and the third, society would become extinct; for the second, life itself would cease; for the fourth, the culture would come to an end. The immense social utility of the nuclear family and the basic reason for its universality thus begin to emerge in strong relief" (ibid.: 10). Foreseeing counter arguments that these functions can be performed by other individual agencies, Murdock asserts: "No society, in short, has succeeded in finding an adequate substitute to the nuclear family, to which it might transfer these functions. It is highly doubtful whether any society ever will succeed in such an attempt" (ibid.: 11).

Another issue is whether the nuclear family does not shed some of its functions and loses its identity when it becomes a constituent unit of a composite structure like an extended family. Linton (1936) makes a distinction between the conjugal family and consanguineous family, the latter revolving around blood relatives, surrounded by a "fringe of unimportant spouses", and on that basis, argues that where there are larger kin-based families, the conjugal dyad loses its importance and the nuclear family loses its identity. But Murdock denies it and reiterates that a nuclear unit consisting of wife, husband, and children retains its identity even as a constituent unit and discharges its functions, though some of them may be worked out in coordination with the larger group. It is on this basis that he claims universal importance for the nuclear family, even though in his sample of 250 societies, only 47 have independent nuclear families.

Endorsing Murdock's view, I may report my own observations of two large families, one in a tribal society and the other in a peasant society. The tribe of Plains Miris lives in the large river island of Majuli in Assam. As I observed them in the late 1950s, they had large extended families. The largest of the families in a village had 64 members. Each family built a long bamboo house on stilts (as a defence against periodical floods). Setting apart the space required for common amenities such as cooking and eating, besides a specially marked area for the head of the family, the rest of the space was divided into a row of compartments with bamboo partitions. A narrow passage running along one side of the structure provided entry to each of them, which was occupied by one constituent nuclear unit – man, wife, and their children. Each nuclear unit slept and stored its exclusive belongings such as clothes, tools, etc. in its allotted space. Until the children came of age, the respective parents were their main caregivers and mentors, and only at a latter stage the youth joined the work teams of elders, cutting across the nuclear units as per the individuals' skills and aptitude. (The choice in that society at that time was mainly between fishing and farming.)

The second case pertained to a landholder's family in Nellore district of Andhra Pradesh. It consisted of about 100 members in the 1940s. Their huge building had, besides common amenities, several rooms and apartments, each of which was exclusively used by a couple and children. The parents were the main socializers for the children up to the latter's adolescence, whereupon they branched off into separate economic sectors such as horticulture, agriculture, poultry, livestock, etc. under the guidance of elders across the nuclear units. So even when nuclear units are constituent parts of a large extended family, they retain their identity and exclusive functions in a large measure. The same kind of set-up could be observed even among some trading communities in urban areas.

Talcot Parsons is one of the prominent sociologists who subscribes to the importance of the three-dyad nuclear family for the human society. After reviewing the functions of the family, Parsons (1955) categorizes two of them as root functions. One of them is socialization of the children and the second is the intertwined development of the personalities of wife and husband in their prolonged partnership. Burgess and his co-authors (1963: 7) present a definition of the family, covering its main features which runs as follows: "The family may be defined as a group of persons united by ties of marriage, blood and adoption, consisting of a single household, interacting and communicating with each other in their respective social roles of husband and wife, mother and father, son and daughter, brother and sister, and creating and maintaining a common culture". These writers, like many others, identify and underline the essential and irreducible components of the family. Without the presence of all these features, there is no complete family for them. Such full-fledged units with the complement of three dyads are referred to in social science literature by alternate terms such as elementary family, conjugal family, or nuclear family. (A few sociologists have tried to make a hair-splitting distinction between nuclear family and conjugal family, but this writer does not see any merit in it.) The term 'nuclear family' carries an inherent suggestion that it is a cell which builds itself into larger units, thus becoming the base for the entire social structure.

II

As against the foregoing view which upholds the universal importance of the nuclear family, there are some social scientists who contend that the conjoint functioning of the three dyads with the simultaneous presence of wife, husband, and children is not essential for the functioning of the society, and the latter can get along with any two or even a single dyad. Among the social scientists of the older generation, Adams figures prominently in projecting the latter view. [Among the contemporary sociologists there are many who do so, and a few such cases will be considered at a later stage.] Adams (1971) considers the functions of the family as mentioned by Murdock and asserts that those very functions can be performed by other individual agencies as well.

His rebuttal is so drastic that it makes one wonder whether he really meant it to be so. One passage in his paper reads as follows: "We simply find that other social agents do in fact take over the functions (of a family) for an extended period of time ... The educational function can be taken care of by the mother, other relations, chums, schools and so on. The rationale that a male child must have a resident father in order to learn to be a man does not hold in fact. The economic functions can be handled by the mother, and the children as they grow older; to this can be added grandparents, brothers and other relatives ... And of course, the sexual function can be handled well by other married men, boarders, visitors, friends and so forth. The reproductive function does not need the fathers' presence; a midwife is more useful" (ibid.: 24–25).

Adams appears to be abundantly confident of the efficacy of the substitutes. What he does not seem to realize is that there is always a difference, a world of difference, between the original and the ersatz. One cannot convincingly argue that the presence of a husband is not necessary for a woman and the presence of a father is not necessary for children. It is as absurd as arguing that children can survive if there are orphanages. When he discusses some cases, he depends on his own observations in central and south America, besides Smith's study of British Guiana which had been through the dark phases of colonial and racial history and which – may be as a result – now have a substantial proportion of woman-headed families. His data reveal that woman-headed households constituted 26 per cent in Nicaragua, 25.5 per cent in Elsalvador, 17.2 per cent in Costa Rica, and 16.8 per cent in Guatemala in the year 1950. From that base, he projects the view that a considerable proportion of population in different societies does not live in nuclear families, but in a domestic set-up which is characterized largely by the maternal dyad, consisting of mother and children, and to a small extent by the paternal dyad, consisting of father and children. According to him, the conjugal dyad appears on the scene but for a short span of time.

Adams confidently asserts that the society is able to move on as the functions supposed to be performed by the nuclear family are discharged by the individual dyads. Therefore, he reaches the conclusion that sociological analysis should proceed not within the framework of the nuclear family but within the framework of the smaller units, namely the dyads. As different units individually discharge these functions it is, according to him, more fruitful to study these units severally. His argument concludes as follows: "For some functions and under some circumstances, the individuals may be effective agents; for others the elemental dyads are more efficient; for yet others, the nuclear family may serve" (ibid.: 30). He rails against the assumption that the absence of nuclear families implies a breakdown of the social system.

His observations do not seem to be fully warranted by the material at his disposal. He himself admits that the nuclear family is the basic unit of society among the native Indians of Guatemala, while among the Ladinos, considerable number of households are headed by women. The East Indian residents of British Guiana retain strong husband-father role in their families. And the author on whom he is supposed to draw, i.e., Smith, is himself shown to hold the view that "the nuclear family does have universality" (ibid.: 25). In a sober mood, Adams himself accepts that "while theoretically, the two kinds of dyads (the maternal and the paternal) can operate independently, the society will be a sadly disappointed affair where they do" (ibid.: 29). Towards the end, he tries to balance between two antithetical positions and sums up as follows: "The assertion that the nuclear family successfully fulfils certain functions is perfectly valid. But the reverse assumption that other social forms can never suitably fulfil these functions is empirically and theoretically invalid" (ibid.: 36).

Adams is wrong in asserting that there are other social forms which can perform the functions of the nuclear family equally effectively. As he himself recognizes three constituent dyads in a family, he should realize that if one of the dyads is not present in a domestic unit, to that extent the unit becomes incomplete or deficient. And an incomplete unit cannot equal a complete unit. Even when other agencies try to cover the gap, the deficiency still remains. There is always a difference between what the interrelated cells of an institution can conjointly perform and what individual agencies can severally perform. The moral code picked up by a child in a Sunday school differs in its effect from the moral code ingrained by a parent who also nurses, tends, and bestows all the necessary care.

But Adams may be right in saying that all the domestic units, which do not have the full complement of the three dyads, should not be considered abnormal or unnatural. When one says that the nuclear family is of universal importance, it does not mean that every individual in every society is living every moment of his life in a nuclear family. This does not happen. The domestic units have their own development cycle, and they experience periodical depletion and replacement. When a man dies or deserts, the family is depleted, as the one who has played the role of husband-father' is lost, and the mother and children are left alone. It becomes a broken family. Looking at things in another way, there may be an young couple without children. In that case, it is not a full-fledged but an incomplete one. When children are born or adopted, it becomes a full-fledged family. So in any society, at a given time, all the domestic units cannot be full-fledged families consisting of all the three dyads.

The question with which we started may be rephrased, i.e., whether we should restrict the term family only to domestic units that consist of all the three dyads, or extend it even to domestic units with two or even a single dyad? My own suggestion is that the term family, technically speaking, should be confined only to domestic units with the full complement of three dyads. Other domestic units of different kinds should be referred to by generic terms as mentioned earlier.

III

There are views to the contrary which must be considered. These are days when we hear shrill voices proclaiming that the days of the nuclear family are numbered, and all the variant and deviant forms should be brought under the rubric of family. In a publication entitled, *Family Diversity*, Pauline Irit Erera (2002) declares on the opening page: "At a dizzying pace, the traditional, two-parent, heterosexual family has given way to a variety of family arrangements". Then, citing other authors such as Hill, she writes: "Today most adults no longer live in a co-resident nuclear family. … the first married, heterosexual family is but one of the numerous family structures" (ibid.: 1). She cites the view of another author to the effect that the children from divorced families turn out to be as good as other children, if all other conditions (including parental resources) are equal (ibid.: 11). The logic of this "if" is noteworthy. Then, she dismisses the idea that households without fathers cannot socialize the male children effectively "as a ploy to retrieve the outmoded sex roles" (ibid.: 12). At a later stage when she has done with her ifs and buts, the harsh reality seems to dawn on her and she concedes that "because of the physical, financial and emotional stress involved in raising children, a family that includes more than one adult is more likely to contribute to positive child outcomes" (ibid.: 208).

Erera takes objection to the omission of lesbians and gays in the mainstream family literature, though according to her information, there are three million of them in the US. Raising her accusing finger, she says that "this omission implies ... that these families are not (considered) real families" (ibid.: 160). Then, feeling less sure of herself, she says "strictly speaking there are no lesbian or gay families, only families with lesbian or gay members. When I refer to lesbian or gay family, this should be understood as a short-hand way of saying families headed by lesbians or gay men" (ibid.: 161). She draws on some studies to show the existence of a lasting relationship among the lesbians and gays but also cites some other studies suggesting a fragile relationship compared to heterosexual couples (ibid.: 171). Again, she refers to one study revealing that the homosexual parenting is as good as heterosexual parenting, followed by a reference to another study which points out that the involvement of the co-parents varies depending on whether the children have been born during an earlier heterosexual relationship or through insemination or they have been adopted (ibid.: 177–78).

The legal position in the US is that most of the states do not permit same-sex marriage. Vermont has a civil unions law which offers same-sex couples the facility of obtaining a certificate. New York and New Mexico permit adoption by lesbians and gays, while some other states like Florida and New Hampshire have taken measures to strictly prohibit it.

Other transformations in the domestic structures in the US

Goode (1965: 101) cites the findings of P.H. Jacobson to the effect that in 1955, there were 3.3 million children aged under 18 years (5.9% of the total population) who lived in single-parent households and there were about 2.7 million orphans. According to Erera (2002: 20), there were 565,000 children in foster homes in 1999 and 5.5 million children resided in households headed by grandparents, which was an increase of 40 per cent since 1980. Erera (ibid.: 101) also reports that while only 9.1 per cent of children under 19 years of age were living in single-parent families in the year 1960, the figure rose to 24.7 per cent by 1992. Elshtain (2000: 2) notes that while 17 per cent of the children were not living with their biological parents in 1960, this category came to represent 36 per cent of the total children under 18 years by the year 2000.

Wilson (1993: 26–27) reports the findings of a survey of 60,000 children living in different kinds of households all over America conducted by the Department of Health and Human Services in 1988. The survey took into consideration the age and sex of the children and the ethnicity, income, and the marital status of the parents. It was found that at every income level, for both the sexes and all ethnic groups alike, children living with a never-married or a divorced mother were substantially worse off than those living in two-parent families. Compared to children living with both the biological parents, children living in single-parent households were twice as likely to have been expelled or suspended from the school. They were also much more likely to display emotional and behavioural problems, and to engage in anti-social behaviour.

According to Elshtain (2000: 2), while out-of-wedlock births constituted 5.3 per cent of the total births in the year 1960, they rose to over 30 per cent by the last decade of the 20th century. Erera (2002: 101) notes that 34 per cent of the single parent families were due to births to unmarried women in 1991. She offers another dimension of the same phenomenon when she says that 37.8 per cent of all births in the year 1994 were to unmarried women (ibid.: 126).

Goode (1965: 94) notes that the divorce rate in America in the year 1956 was 246 (for 1,000 marriages). Levitan and Belous (1981: 12) estimate the divorce rate to be two in five marriages, i.e., 400 for 1,000 marriages in the 1970s. Erera (2002: 203) makes a resounding statement on the divorce situation in America. She says: "Half of all the married women who divorce, along with those women who do not marry, make a powerful statement: that marriage and traditional family life are oppressive for them … Indeed, two out of every three divorces are now initiated by women". According to Hacker (1982: 38), while 10.2 per cent couples between 18 and 34 years were without children in 1970, this proportion rose to 17.6 per cent in 1980. Giddens (2004: 58) says that 25 per cent of couples in the age group of 18–35 years in the western countries do not want to have any children.

Levitan and Belous (1981: 8) note that towards the end of 1970, divorced, separated, widowed, and never married women headed more than eight million households, or one in every seven in the US. In addition, there were childless couples and those who lived alone. They conclude that "the living arrangements of more than 75 per cent of the population differ from normal family", while the facts they report are unsavoury; curiously they try to cover it with the veneer of growing pluralism in American society. "These arrangements", they claim, "should be viewed as just, as legitimate, as traditional household model". This provokes an apt response from Hacker (1982: 38) which reads as follows: "This is a curious term of description. Instead of the conventional family type composed of wife, husband and children, we now have pluralism. Well perhaps. For these writers the pluralistic family could include a mother struggling to raise her kids, without the father or his income." Relevant in this context is Erera's, admission that "nearly half of single parent families live in poverty" (2002: 100).

Some crucial dimensions

In his review essay with a tell-tale title, *Farewell to Family*, Hacker (1982: 37–38) lists a few crucial dimensions of the deviant domestic structures. The American census classifies households under three heads: family households, non-family households, and single-member households. A family household is defined as one in which any two members are related to each other. (This may itself be a wrong use of the term family, but we may leave it at this point.) A non-family household is one in which no two members are related to each other. Our main concern here is about the galloping growth of this category of households in America in the recent decades. Taking all these types of households together, they rose from 63,401,000 to 79,108,000 between 1970 and

1980 – a growth of 24.8 per cent as against 11.5 per cent increase in population. This disproportionate rise in the number of households was due to a spurt in the growth of non-family households. Of the 15,707,000 freshly added households, as many as 8,763,000 were non-family households. It meant that among the total number of freshly added households, the non-family households constituted 55.6 per cent. From 11,919,000 in 1970, the non-family households reached the figure of 20,682,000 by 1980, recording a growth rate of 73 per cent in a decade. A break-up of the non-family households showed that 25.2 per cent consisted of two men, 19.2 per cent consisted of two women, and 55.6 per cent were of mixed sex. Among the additions to family households, 3,518,000 were single-parent households and 3,452,000 were married couples. It meant that only 22 per cent of the total household growth came from married couples. Hacker also mentions that single-parent households grew by 52.3 per cent between 1970 and 1980.

As could be seen from the foregoing account, many of the units included in the category of family households are not full-fledged families. A large proportion of them are single-dyad units. According to Erera (2002: 1–3), in the year 1950, about 60 per cent of the American households conformed to the nuclear family structure. By 1970, about 45 per cent of the households were composed of married couples with children. By 1996, only 26 per cent of the households corresponded to that category. This should be enough to startle any social thinker. More than the gross figures, the trends should be a cause of worry for all those concerned with family values. But a number of sociologists, for different reasons, are justifying and defending these trends.

Wilson (1993: 25) narrates a debate between some sociologists on the relevance of two-parent family. During the autumn of 1992, the views of two sociologists disclaiming the importance of the role of father in a family, were published in the *Washington Post*. This evoked a rejoinder from Professor Propenoe of Rutgers University, which was published in the *New York Times*. Therein, he affirmed that both mother and father have important roles in running the home and in socializing the children. Immediately, another academic, Judith Stay joined the fray accusing Propenoe of propagating a widespread prejudice. Another sociologist, Jessie Bernard went hammer and thongs at the very idea of treating marriage as a privileged institution, while there are many alternatives to it. A few sociologists did support Propenoe, affirming the desirability of two-parent family, but cautioned him, at the same time, not to offend the sentiments of those large numbers of persons living in single-parent household. They should take a leaf from Obama's image of his father. Barack Obama was a child when his African father and white mother separated. Then, he spent a few years in Indonesia with his stepfather before moving back to America to live with his grand-parents and complete his school education. Speaking of his grandfather as a man of disappointments and of his stepfather as a man of compromises, he contrasts them with his father whom he refers to as a man of his dreams. His edifying words read as follows: "These men (the former two) become object lessons for me, men I might love but never emulate, white man, brown man whose fates did not speak to my own. It was into my father's image, the black man, the son of Africa, that I had placed all the attributes I sought in myself" (2007: 220).

Here is the crux of the matter. When a social trend is deleterious, what should social scientists do. When you tell a delinquent that he is in need of reform, he may feel hurt. For that reason, should a sociologist refrain from suggesting reform? Of course in the domestic sphere the children are often the victims of circumstances beyond their control. But some reforms should be attempted at the level of adolescents and adults so that the trends can be arrested and even reversed. The social scientists have a role and responsibility in the matter. They cannot say that there is nothing like a norm, and any deviation is welcome as an addition to pluralism. If norm and deviation are not distinguished from each other, there is no sociology.

IV

The foregoing account pertains to America. The picture is almost the same in all the western countries, which are supposed to set the trends for the rest of the world. In this context, one question needs to be posed looking at things from the other end. Are not some of the features mentioned in the preceding pages present in simpler societies? The answer is yes; some of them do exist. There are primitive societies among whom, the divorce rates are as high or even higher. But there is a difference. There, the divorcees remarry, and the majority of young adults of both sexes at any time are in a wedded state, playing the roles of spouses and discharging the duties of parents. There may be broken homes in simpler societies, but the majority of people strive to live in nuclear or extended families with the three dyads intact. The aim and the ideal are the essence and they make the difference.

One exception which occurs in a few matrilineal societies needs to be mentioned and explained. Many of the matrilineal societies follow the uxorilocal rule, according to which the husband moves to the wife's village and sets up a home. When children are born, they form a nuclear family. But there are a few matrilineal societies which practice natolocal rule of residence, according to which, brothers and sisters continue to live together in the household where they have been born, with the eldest male member playing the role of the head of the household, and the sisters receiving their husbands at night. During the day, the men discharge their duties towards their sisters and the latter's children in their own homes. This is a rare feature observed in a few societies like the Nayars of Kerala, the Muslim inhabitants of the Lakshadweep islands, the Pnar section of the Khasis in Meghalaya, and the Ashanti of Africa. In the first three of these societies, there is a system of visiting husbands and in the fourth the wives visit their husbands' homes at night. Most of them contract intra-village marriages. (Murdock was mistaken in denying this cultural practice and in challenging the authenticity of the Nayar case cited by Linton.) But what must be noted in respect of these cases is that their households run on well-established lines with brothers and sisters playing their traditionally determined roles in managing the domestic economy and in bringing up the children. And there is a socially approved marital bond between a man and his wife, and a culturally recognized fatherhood. So, there are three dyads present and operative in these societies, though not in the ideal matrix of a nuclear family. It is the same with societies which are matrilineal and avanculocal, though in their case the three dyads operate in a more cohesive matrix. But in the large number of cases of single-parent households and non-family households in America, there is a deficit or even a total absence of dyads.

These days we hear voices among social scientists holding forth that there is no justification in privileging family or marriage. Jessie Bernard who wrote a book entitled *Future of Marriage* asserts that there is nothing sacrosanct about monogamy and reels of several alternatives which are open to individuals to choose according to their whims and convenience. The alternatives she spells out are: group marriage, commune living, marital swinging, cohabitation, limited commitment marriage, etc. (cited by Wilson, 1993: 25). Writers like her seem to be carried away by some evolutionist theories postulating that man at the beginning practised promiscuity or group marriage, and did not have anything like a family. That is a mistaken view. A proto-type of family exists even among the primates. One should heed what Edward Tylor, the doyen of the exponents of the evolutionist theory of culture, has to say on this matter. Tylor (1960: 246) says: "Mankind could never have lived as a mere struggling crowd, each for himself. Society is always made up of families or households bound together by kindly ties, controlled by rules of marriage and the duties of parent and child".

It may be appropriate too, in this context, to recall the views of some of the leading sociologists upholding the importance of the family for man and society. According to Durkheim, the family is the basic link in the chain of social being and it is the cornerstone of social structure. Brigitte Berger and Peter Berger (1983: 174) say that only individuals produced by stable families can be the foundation for a stable society and stable democracy. More importantly, they point out that "the family remains the institution in which, at any rate, the very great majority of individuals learn, whatever they learn, about morality" (ibid.: 176). Goode (1965: 109) speaks of the benefit that flows from the family even in industrialized societies. He says: "Without the family to deal with the idiosyncrasies of the aged, the emotional needs of the adults, and the insecurities of the children, very likely, not enough adequately functioning people would be produced to man the industrial system". It is relevant to recall the high pedestal the family occupied in the soviet society and the way it served as a shock-absorber against the stresses and strains of a regimented society.

It is apt to listen to the sage counsel of Whitehead (1993: 77) regarding the distortions in the family structures. It runs as follows: "Family diversity in the form of single-parent and step parent families ... dramatically wakens and undermines society ... The new forms are not an improvement on the nuclear family, nor are they as good, whether you look at the outcomes for children or outcomes for the society as a whole". Lastly, we may turn to the expression of angst by a black lady activist, Eleanor Norton at the deteriorating family life even a few decades ago. Regretting the fast growth of women-headed households and the rise of premarital births to teens among the blacks, she draws attention to the existence of parallels among the working class whites, which, according to her, are qualitatively the same, though differing in degree. We may read her in her own words: "The repair of the Black condition in America depends on the succour of strong families ... We must make marriage and family a tool for improving our lives ... No institution is more beloved in black consciousness; none (else) is at the heart of social hope" (cited by Novak, 1976: 8).

Giddens (2004) sees many of the current changes in the domestic life as the inevitable results of the ubiquitous socio-economic forces. Yet, he sees some signs of waking up to the seriousness of the situation in China as well as in the western countries. He refers to the Chinese government taking measures to make divorce more difficult and to relax the one-child norm. "In China", he says "there is much talk of protecting the family. In many western countries the debate is even more shrill. Politicians and activists routinely diagnose the break-down of family life and call for a return to the traditional family" (ibid.: 53). Giddens emphasizes the importance of trust among members of the family which, according to him, springs from "a democracy of emotions" within the family. Above all, he lays store by the resilience of the family and its capacity to survive crises. Let us hope that his hope is well founded.

No one should assume that a call to save the family is a call for gender inequality or for hierarchical age grades among its members. Nor should it imply the shutting of the door to unavoidable divorces. A reasonable ideal that can be projected is a three-dyad family with gender equality and context-related give-and-take between husband and wife, parent and child, and among siblings, based on what Giddens calls democracy of emotions. The overall aim must be to make the family less fragile and more stable. Deviations do occur; they cannot be ruled out. But, they should be within limits such that the deviant does not supplant the archetypal.

The current trends threatening the future of the family should serve as a wake-up call for social thinkers and statesmen alike. The family is intimately interlinked with society and culture, nay, with the human nature itself. It is the crucible in which the future generations are shaped. Without families governed by well-established norms, there is no coherent society. And without the latter, man does not remain a human.

References

Adams, R.N. 1971. The Nature of the Family. In J. Goody (ed.) *Kinship*. Baltimore: Penguin Books.

Berger, B. and Berger, P. 1983. *The War Over the Family*. New York: Anchor Books/Doubleday.

Burgess, E.W. et al. 1963. *The Family*. New York: Van Norstrand Reinhold.

Elshtain, J.B. 2000. Reflections on the family at the millennium's beginning. *World and I*, March 2000.

Erera, P.I. 2002. *Family Diversity*. London/New Delhi: Sage.

Giddens, A. 2004. *Runaway World*. London: Profile Books.

Goode, W. 1965. *The Family*. New Delhi: Prentice Hall of India.

Hacker, A. 1982. *Farewell to the family*. New York Review of Books, March, 18: 37–44.

Levitan, S.A. and Belous, R.S. 1981. *What is Happening to the American Family*. Baltimore: The Johns Hopkins University Press.

Linton, R. 1936. *The Study of Man*. New York: Appleton–Century–Crafts.

Murdock, P. 1960. *Social Structure*. New York: McMillan.

Novak, M. 1976. Family out of Favour. *Harper's* 252 (April): 37–46.

Obama, B. 2007. *Dreams from My Father*. Edinburgh: Canongate.

Parsons, T. 1955. The American family: its relations to personality and to the social structure. In T. Parsons and R.P. Bales (eds) *Family, Socialization and Interaction*. New York: Free Press.

Tylor, E.D. 1960. *Anthropology*. Ann Arbor: University of Michigan Press.

Whitehead, B.D. 1996. Dan Quale was right. *Atlantic Monthly* 274(4): 47.

Wilson, J. 1993. The family values debate. *Commentary* April 1993: 24.

7 Race and Ethnicity

A Methodological Note with Reference to the Study of Indian Diaspora

Ravindra K. Jain

One of the persistent, emic*, preconceptions in most cultural communities of the world has been their self-conception as a "we group". Often it is stated in terms of equating the community with humanity at large. As analysts of human society and culture, both in their particularities and generality, we are bound methodologically to take into account this "emic" preoccupation for after all the necessity of a dialogue (the so-called "dialogical" approach in ethnography and social anthropology) characterizes the interface between the observer and the observed. On the other hand, while incorporating these self-perceptions (emic data or what Levi-Strauss, 1963, called "conscious models") in

*According to Wikipedia, "an 'emic' account is a description of behaviour or a belief in terms meaningful (consciously or unconsciously) to the actor; that is, an emic account comes from a person within the culture. An 'etic' account is a description of a behaviour or a belief by an observer, in terms that can be applied to other cultures; that is, an etic account is 'culturally neutral'."

our analyses and explanations, we have to also go beyond. In the latter procedure, we may not only have to encompass the conscious models into "unconscious" ones to use the Levi-Straussian terminology, but also attempt "destructive analysis" (Handler, 1985) of emic concepts, particularly in circumstances where the actual isomorphism between the people's models and those of the analyst poses an insurmountable obstacle to an explanation of the phenomena in question. To put it in the framework of what Oxford anthropologists termed as "cultural translation" (cf. Evans-Pritchard, 1962; Lienhardt, 1954), the task of converting particular differences in cultural meanings into general propositions (while retaining the particular nuances of meaning across cultures) could be accomplished only when the emic is surmounted by the "etic" while at the same time, preserving the essence of emic meanings and, indeed, augmenting the former by nuances of the latter.

In what follows, I am going to argue (after Handler) that while it would be grossly counterproductive for the anthropological analyst to stop short at the convergence of words, tropes, and meanings between the observer and the observed, it should be underscored (a proposition Handler doesn't make) that the "destructive analysis of the familiar" in anthropological methodology must guard against throwing the baby away with the bathwater, viz., fail to retain what is of value in the convergent tropes of cultural commonsense between the observer and the observed. An example of the latter is what I alluded to earlier as the universal community commonsensical self-perception of being a "we group", almost unique and sometimes synonymous with humanity at large, also called 'ethnocentrism' by anthropologists. Bringing this realization closer to a methodological perspective that we wish to advance here, a certain sub-stratum of a belief in "social heredity" underlies many a community self-perception (Jain, 1996). The objection that this proposition is essentialist in nature may be offset by its analytical gains. In what follows, I wish to demonstrate precisely this feature – that our proposition underlies nearly universal beliefs of relationship between difference and hierarchy – with limited reference to the semantic continuum between ethnicity and race (and implicitly 'class') as analytical concepts. The comparative ethnographic evidence is provided from the field of Indian diaspora which has preoccupied my research attention for nearly the last 50 years.

Before proceeding with my demonstration of the analytical utility of positing a semantic continuum between race and ethnicity in the comparative study of Indian diaspora, I would like to make a further methodological observation. In fact, the observation bears on two related modalities. One, that in the study of socio-cultural phenomena, we must read ontology and epistemology on the same page. In simplified terms, the spatio-temporal generation of systems, structures, and events in our narratives (ontology) has to be seen as shadowed by the intellectual means of comprehending it (epistemology). In other words, to put the issue in the framework of analytical philosophy, in a scientific endeavour of our sort, the ethnographer-analyst does not simply "abstract" general principles from "raw data" (as in most versions of positivistic natural science procedures) but tries to establish what Ludwig Wittgenstein termed as "family resemblances" in the interplay between the ontology and epistemology in the study of phenomena (Wittgenstein, 1953) (see in this respect the anthropologist R. Needham's, 1975, methodology of polythetic classification in studying the phenomenon of "kinship"). The procedure is salutary also from the viewpoint of avoiding reification of concepts.

Race and ethnicity

It should be easy enough to show that both the notions of race and ethnicity are refractions of the idea of social heredity. In this respect, the *popular* notion of race rests most blatantly on the phenotype of colour and other physical features of various shades and modalities. Its authoritative scientific distillation has been given early (Metraux, 1950). In 1950, the UNESCO statement, "The Race Question" signed by some internationally reputed scholars of the time suggested that "national, religious, geographic, linguistic and cultural groups do not necessarily coincide with racial groups: and the cultural traits of such groups have no demonstrated genetic connection with racial traits. Because serious errors of this kind are habitually committed when the term 'race' is used in popular parlance, it would be better when speaking of human races to drop the term 'race' altogether and speak of 'ethnic groups'". Despite these scientific warnings, race as "racism" continues to share a conjoint ontological and epistemological basis. On the one hand, its historical connections with colonialism and imperialism have been amply demonstrated. On the other hand, in popular imagination, the idea of race also undergirds current theories of development and underdevelopment, and of hierarchical relations between the "self" and the "other". The latter mindset provides explanations not only of the superiority of the North over the South or of the West over the East as in Orientalist discourses; even its inverse figures prominently in rabid cultural nationalist discourses. This popular concept of race may be distinguished from that of ethnicity which does not always involve a phenotypical biological basis for difference and hierarchy and unlike "race relations" is not always based on strong antagonistic relations between phenotypical biologically defined groups (Rex, 1973). There is, however, an overlap between the two concepts both historically and in sociological analysis as we shall presently see.

In relation to the study of the Indian diaspora, as I have shown elsewhere (Jain, 1989), the universe of discourse in the Caribbean 'in general' locates Trinidadians of Indian origin (Indo-Trinidadians or East Indians) in a racial framework, while the social scientific analysis of Indian Malaysians has been done in terms of a multi-ethnic society. Obviously, the absence of a strong signifier of "colour" in the latter case and its presence in many shades in the former seems to have been the main deciding factor. However, in analyses where the class and power dimensions are explicitly invoked in relation to ethnic and racial differences and hierarchy, it is the continuum between the two notions which takes precedence (see, Stenson, 1980 for foregrounding race and Rasiah 1999 for emphasizing class in Malaysia; see also Alexander, 1977 for colour and class in Jamaica). For analytical purposes, then, there can be no solecism in using either ethnicity or race in relation to difference and hierarchy in particular geographical locations. Indeed, I would argue that if ideas of social heredity seem to be the common denominator, then it behooves sociological and anthropological analyses of particular instances to juxtapose notions, not only of race and ethnicity, but also of class, caste, and culture in the spectrum of concepts deployed for delineating relationships of difference and hierarchy among groups. It can be shown in detail how the use of basically "cultural" paradigms such as "plural society" and "multi-ethnicity" in the Malaysian context has served to exclude the power dimension in emerging social processes (cf. Jain, 2010; also Jayawardena, 1980 for Guyana and Fiji). Similarly, the analysis of British and French societies, as being exclusively based on socio-economic class structure, had lead to the neglect of a cultural dimension, viz., "taste" (Bourdieu, 1984), "blue blood" considerations in defining the aristocracy in UK, and what Geoffrey Gorer (1957) described as the perils of "male hypergamy" in the English class structure. (These are all instances of what we have called "social heredity" in the belief systems of the so-called advanced western societies.) The acrimonious contemporary debate on whether or not "caste" in India is the same as "race" elsewhere reveals once more the solecism in the use of concepts that bear a "family resemblance" to one another. In this respect, the deployment of the concept "ethnicity" as a processual, boundary-marking-and-maintaining (rather than a corporate "group" concept) marks a significant advance in the penumbra-like or sliding-scale use of sociological concepts (cf. Barth, 1969; Brubaker, 2004; Shamsul, 2008; Weber, 1978).

There is, however, an analytical advantage in following through, at least ontologically, on which side of the continuum and of what range and durability a particular system of groups based on social heredity is inflected. This would depend on the ethnographic perspicacity of the anthropologist on the one hand, and, to reiterate, a readiness to read the ontological and epistemological aspects of the phenomenon on a single page, on the other hand. Thus, in the changing ontological scenario of several younger generations of "caste"-oriented Indian communities having been born and brought up in the "race"-oriented milieu of Trinidad or South Africa, the former "passes" into the latter in emic and etic terms (Jain, 2010a). Further, this is an ongoing process. The second-and third-generation Indian youth in post-Obama USA show an uneasy co-existence of their "referent origin" (Dufoix, 2008) identities and identifications, those based on region, caste, language, and religion, with those of class and race in the "host" society (Shah, 2009). Further, there is a historical point to be noted in the slippage between race and ethnicity in sociological analysis. The notion of ethnicity emerged in the USA to designate the non-WASP (White Anglo-Saxon Protestant) *Whites*, immigrants mainly from Eastern and Central Europe, to distinguish them from the racially different Blacks (Kymlicka, 1989). No wonder that even today in the sociological analysis of the US plural society ethnicity is regarded as an "inclusive" and race as an "exclusive" conceptual category (Sanjek, 1994). Furthermore, there has been a tendency for this US-specific historical connotation of "inclusive" ethnicity to diffuse in the analysis of disparate situations, viz., the lingering usage of multi-ethnicity *sans* the exclusion based on hierarchical power relations between the majority and minorities in Malaysia (cf. Jain, 2010). This would show how the application of ethnicity as an analytical tool elsewhere (e.g., in Malaysia) can carry with it the value-load of racist contamination of its founding application.

References

Alexander, J. 1977. The culture of race in middle-class Kingston, Jamaica. *Am. Ethnol.* 4: 413–435.

Barth, F. 1969. Introduction. In F. Barth (ed.) *Ethnic Groups and Boundaries: The Social Organization of Cultural Difference*. pp. 9–38. London: Allen & Unwin.

Bourdieu, P. 1984. *Distinction*. London: Routledge and Kegan Paul.

Brubaker, R. 2004. *Ethnicity without Groups*. Cambridge: Harvard University Press.

Dufoix, S. 2008. *Diasporas*. Translated by William Rodarmor. Berkeley, CA: University of California Press.

Evans-Pritchard, E.E. 1962. *Social Anthropology and Other Essays*. New York: The Free Press.

Gorer, G. 1957. The perils of hypergamy. *New Statesman,* 4th May: 566–568.

Handler, R. 1985. On dialogue and destructive analysis: problems in narrating nationalism and ethnicity. *J. Anthropol. Res.* 41(2): 171–182.

Jain, R.K. 1989. Race relations, ethnicity, class and culture: a comparison of Indians in Trinidad and Malaysia. *Sociol. Bull.* 38(1): 57–69.

Jain, R.K. 1996. Hierarchy, hegemony and dominance: politics of ethnicity in Uttar Pradesh, 1995. *Econ. Polit. Wkly.* 31(4): 215–223.

Jain, R.K. 2010. *Plural Society, Ethnicity, Class: Malaysian Indians and the Crisis of Political Economy.* Bangi: University Kebangsaan Malaysia.

Jain, R.K. 2010a. *Nation, Diaspora, Trans-Nation: Reflections from India.* New Delhi/ London/New York: Routledge.

Jayawardena, C. 1980. Culture and Ethnicity in Guyana and Fiji. *Man* (n.s.) 15: 430–450.

Kymlicka, W. 1989. *Liberalism, Community, Culture*. Oxford: Oxford University Press.

Levi-Strauss, C. 1963. *Structural Anthropology*. New York/ London: Basic Books.

Lienhardt, G. 1954. Modes of Thought. In E.E. Evans-Pritchard et al. (eds) *The Institutions of Primitive Society*. pp. 95–107. Oxford: Basil Blackwell.

Metraux, A. 1950. United Nations Economic and Security Council statement by experts on problems of race. *Am. Anthropol.* 53(1): 142–145.

Needham, R. 1975. Polythetic classification: convergence and correspondences. *Man* (n.s.) 10: 349–369.

Rasiah, R. 1999. Class, ethnicity and economic development in Malaysia. In G. Rodan et al. (eds) *The Political Economy of South-East Asia: An Introduction.* pp. 121–147. Melbourne: Oxford University Press.

Rex, J. 1973. *Race, Colonialism and the City*. London: Routledge and Kegan Paul.

Sanjek, R. 1994. Intermarriage and the future of races in the United States. In S. Gregory and R. Sanjek (eds) *Race*. New Jersey: Rutgers University Press.

Shah, S.P. 2009. Making sense of the second generation. *Econ. Polit. Wkly.* 64(30): 14–17.

Shamsul, A.B. 2008. *Many Ethnicities, Many Cultures, One Nation: The Malaysian Experience*. Bangi: UKM.

Stenson, M. 1980. *Class, Race and Colonialism in West Malaysia: The Indian Case*. St. Lucia: University of Queensland Press.

Weber, M. 1978. *Economy and Society: An Outline of Interpretive Sociology* (2 volumes) Guenther Roth and Claus Wittich (eds). Berkeley: University of California Press (first printing 1968).

Wittgenstein, L. 1953. *Philosophical Investigations*. Oxford: Blackwell.

8 Colonialism-in-the-Mind

Gouranga P. Chattopadhyay

While discussing the phenomenon of celebrating numerous useless "educative" acts like loading young brains with such data as the number of brats brought to earth by the "prim" queen Victoria, Dr Ashok Mitra (*The Telegraph*: 30 March 2009) states that it reflects the moneyed Indian class's attraction towards the West that has opened the door for them to fill their coffers. He calls this phenomenon a kind of colonialism.

While I agree with Dr Mitra's hypothesis, my intention in this article is to show that this hangover of colonialism runs far deeper in the Indian psyche. It is difficult to realize this continued hangover of colonialism most likely because it has been lodged in the collective unconscious. (I use the words "collective unconscious" not in the Jungian sense, but in the way that Hopper has used it in the recent years.)

To that end, first, I shall examine some of the overt phenomena that are part and parcel of life in India today.

One such noteworthy phenomenon is the scramble for "good" schools that provide education in English medium in exchange of what at times amounts to "a King's" ransom, as opposed to other schools that use "regional" languages as the medium of instruction. Please note the words "regional language". I was born in British India, fortunately in a family of freedom fighters. Along with the rule at home of buying every thing Indian as opposed to imported foreign goods, I went to a Bengali-medium school. In those days, the medium of education in every Indian language was considered as getting educated in one's mother tongue of that region and not in a "regional language". We knew that our career in future was likely to suffer irrespective of the quality of English language that we learnt because of the very fact that we did not go to an English-medium school. A strong sense of commitment to our mother tongue allowed us to deal with that anxiety. The British imperialists had tried their best to destroy this commitment and undermine the status of local languages by using the term "vernacular" to describe the various languages that Indians used as mother tongue. The Oxford Dictionaries of the past defined the word *vernacular* as "pertaining to the tongue of the slave", i.e., the Indians were slaves in the minds of the British as part of the US and European heritage of actual slave trade that was continued in the garb of indentured labour in the colonies till well after slave trade was abolished. So, their languages were described as the language of the slaves.

Much later, after the Indian empire vanished and in its place there came the nations of India, Pakistan, Sri Lanka and Myanmar, the definition of "vernacular" also got changed in the later editions of *Oxford Dictionary*. The new definition became the language or dialect spoken by the ordinary people of a country or region. It leaves me to wonder if in the mind of the British, the status of being colonial masters still exists, so that while the "ordinary people" speak "vernacular", some kind of extraordinary status is conferred upon those people in former colonies who speak English.

This emphasis on "ordinary people" seems to have impacted the Indian psyche to produce the idea that students of English-medium schools belong to an "exclusive" class of people who are perhaps a cut above the "ordinary people" who can afford education *only in the "regional language"*.

So the "colonial hangover" is about being politically correct in raising the status of "slave" to the "ordinary", the hoi polloi. The tragedy of the country is that the average hoi polloi has also accepted this constructed "reality" and continues to use the word "vernacular" to denote various local Indian languages.

Another very destructive ploy of the British imperialists was translating the word *dharma* into English as "religion". Very few people seem to have given much thought to this misnomer. *Dharma* in the past, was a very inclusive term with multiple meanings. Some of these meanings still remain. At one level it is applied to a very wide range of realities called *monushya dharma* to highlight the common human inner realities and behavioural characteristics as opposed to those of animals (or shall we say various species of animals other than the species termed as *Homo sapiens*). The definition of *dharma* then keeps narrowing its boundary through such definitions as the expected ethics and behaviour of kings and other varieties of rulers "by divine right" (*raj dharma*) that were accepted and got incorporated in the *dharma* enforced by the powerful Vedic people (the *brahmanya dharma*), down to the personal philosophy and ethics of creative people who questioned the prevalent culture.

Such wide definition of *dharma* created the ground for tolerance and dialogues across people with different philosophies, faiths, and beliefs. That did not suit the British imperialists, who depended on a policy of divide and rule. By destroying the earlier education systems prevalent in this subcontinent that had an important component of experiential learning at individual and group levels, they introduced a kind of pseudo-western education system. Its thrust was to produce clerks and various lower-level employees. Destruction of the meaning of *dharma* and changing it into a very divisive European concept called religion was an important tool for executing the power of divide and rule. Since experiential learning is a very powerful tool for locating one's truth within oneself that leads to questioning many cultural, including politico-economic, assumptions, the imperialists also took care to define the word *darshan* as philosophy. However, since *darshan* literally means "to see" or "seeing", use of that word points to the fact that behavioural prescriptions and proscriptions arose from experience-based consensus. But once *darshan* was translated as philosophy and this new meaning was taught at all levels of education, like in the case of *dharma*, the idea of experiential learning gradually vanished from the subcontinent till, sadly (for me at least), it was once again learnt from the USA and England. I am here referring to t-groups or sensitivity training started in the USA, which reduces all relationships to interpersonal, thus knocking out the possibility of questioning societal and organizational assumptions, and group relations conference (also known as working conference), introduced by Tavistock Institute of UK that keeps it focus on inter- and intra-group assumptions from very small groups to larger and larger groups. It is interesting to note that this latter form of experiential learning is based on the work of W.R. Bion, who was born in India and imbibed many Indian ideas and notions unconsciously. He got in touch with this inner reality and acknowledged it rather late in life and decided to revisit the subcontinent. Unfortunately, he died before he could undertake the trip.

Division of the Indian empire of the British in South East Asia into Sri Lanka and Myanmar was based on political and economic reasons. But, the other division – that in to India and Pakistan – was based largely on religion. To be precise, this division was based on Mohammad Ali Jinnah's two-nation theory, i.e., that the Hindus and Moslems comprised of two different nations. To reinforce this constructed reality of two-nation theory, this part of the division was named in English as Partition, while partitioning of erstwhile Burma as Myanmar and Ceylon as Sri Lanka were not partition in English language.

Sigmund Freud used the term *defence mechanism* to include a number of processes that people employ to unconsciously deny or distort an experienced reality. This denial or distortion does not permanently change one's inner reality. It remains intact in the unconscious and gets acted out, also unconsciously. One of those inner realities is identifying with the aggressor unconsciously to avoid punishment and curry favour.

In British India, for almost two centuries, this process had taken place. A very small percentage of people, variously termed by the British as "terrorists" and "enemies of the king or the queen", depending on the gender of the person who sat on the throne in London at that time, resisted the fear of punishment, accepted the punishments meted out by the imperial rulers, and kept up the fight for freedom that culminated in the

Union Jack going down at the hour of midnight on August 15, 1947 and the tricolour with Chakra in the centre piece going up.

Sadly, the new Indian government could not gauge the depth of penetration of identification with the aggressor in the psyche of the average Indian – both the so-called "ordinary" people and the "exclusive" people. As a result, no steps were taken to re-mould the education system, re-establish some of the relevant pedagogical methodologies followed in pre-British India, and freeing the Indian psyche from the internalized identification with the aggressor as a metacultural reality in India (and presumably in Pakistan too).

That is the colonial hangover. It has largely alienated Indians from many other realities that are part of the subcontinental metaculture that are not internalized. Those metacultural realities have the creative power necessary to establish a *Mahan Bharat*, a Great India. Instead the desire for greatness is projected on the rear bumpers of public busses and trucks by writing *Mera Bharat Mahan* (my India is great), while individuals keep behaving mostly as something which is opposite of *mahan* (great).

In the early 1940s, as a school student, I had attended an annual conference of All Indian Student Federation in Guntur (which was then a small village). During that conference, some of the local members had sung a song in Telegu, the first line of which I still remember: *Desha mante mattikadoi desha mante manusuoi* (The country does not consist of its earth only, but it includes the human beings too). This *Mera Bharat Mahan* legend written in various places seems to push the onus of being *mahan* (great) on to the earth of India, while releasing the people to be mean and to behave like adolescents to downright infants.

There are other evidences of the colonial hangover in various aspects of life. One of them is some kind of a belief about "centre-state relationship" that has not been consciously articulated, and therefore unexamined about some of its deeper implications. Here "centre" is represented by the national government with its seat in Delhi (New Delhi to be accurate). In reality this government headed by a Prime Minister and his cabinet, is authorized by the adult population of India. This is because the process is one of election to the Lok Sabha (the Lower House of the parliament) based on adult franchise every five years (unless the mandate gets withdrawn through various constitutional processes earlier). There is also the Rajya Sabha (Upper House) with limited authority. The President of the Indian Republic, who is authorized to hold that office by the people's representatives to the Lok Sabha and the Rajya Sabha, administers the oath of office. Then, there are the various state governments that are also headed by people's representatives through election. So, since both the central and the state governments are run by the representatives of the adult population of India, and the Republic's constitution spells out the nature of centre-state relationship, any dispute or difference occurring between the centre and one or more states logically requires resolution through constitutional process that also allows legal action.

However, there are all too many instances of the political parties, including the one in power in a state government, to give call for general strike against certain decisions taken at the centre that affect the state. Only, such general strikes are called *bandh* (literally closure or a dam) and in character have little or no difference from general strikes. This usually happens when the party in power in a state sits in the opposition bench in the parliament. It then becomes an absurd situation of opposing a decision taken by people's representatives at the centre by a political party that controls the state government. To that extent, the state government covertly supports the *bandh.* I am calling it absurd because it becomes a fight between two sets of office holders, both of whom derive their authority from the people of India. My hypothesis to explain this absurd phenomenon is that this too is a colonial hangover. Delhi has been the seat of power held by foreigners from 1206 when the Turks conquered parts of northern India and started ruling from Delhi. This process ended only on August 15, 1947, when the British had to part with their Indian empire that got divided in to present day India (Bharat), Pakistan, Bangladesh (erstwhile East Pakistan), Myanmar (erstwhile Burma), and Sri Lanka (erstwhile Ceylon).

In the unconscious of the Indian psyche, it seems that the central government still remains as some kind of "foreign" power that gives "legitimacy" to the conscious part of the psyche of the dominant political parties that form the state governments to call *bandhs*. The parties in the opposition in different states too then follow this model and organize *bandhs* to oppose the decisions taken by their respective state governments. That such *bandhs* are destructive for the country's economy or that it is the poor – the daily wage earners, small shop keepers, and other small businesses, as also those who deal in such products as vegetables, fish, fruits etc. that are sold daily in municipal markets – who are the worst sufferers, seems to be of no concern to the *bandh* organizers. It is as though like in the colonial days, political parties assume that great sacrifices are required by the people to shake off foreign imperial power! The adolescent aspect of these *bandhs* is indiscriminate vandalization of both private and public property to express one's anger against the powers that be. Once again, it is as though state-owned property is not created out of people's money, but from the ill-gained money of some foreign imperial power.

This seems to be truer in West Bengal than in other states because, I further hypothesis, the seat of British power in India was originally Bengal with Calcutta as the capital of British India. Much later it was shifted to Delhi (New Delhi). So, the psyche of the Bengalis hold the idea of power being taken away by those who rule from Delhi and therefore *bandhs* are called in West Bengal almost at the drop of a hat!

It is also well known that little or no change has been done in the manuals of various all-India civil services such as the IAS, the IPS etc. The IAS academy in Mussoorie continues to start life in the morning by giving riding lesson, as though IAS officers posted in India's vast rural areas even today require to ride horses like their predecessors – the ICS – in the absence of motorable roads, like in British India. While the British police academy today have a training schedule of 9 am to 5 pm for five days of the week, the police academy in Hyderabad runs a punishing schedule of 6 am to 6 pm, seven days a week. They also continue to follow the schedule of alternate week of indoor lectures and outdoor physical exercise of various kinds. I was told by some senior police officers that this schedule is followed even though at the beginning of every week of indoors work many participants fall asleep from sheer exhaustion. Further, when the probation period ends, the load of paper work, also a legacy of the British administration, is such that the IPS officers get little time for physical exercise. After the punishing physical activities of the academy, this leads to fat accumulation among young IPS officers.

Being a foreign power, the British did not trust the colonized people. Therefore, many checks were introduced, mostly unnecessarily. This practice also continues today. This leads to a huge amount of paper work, in addition to every government form (and those of semi-government organizations as well) asking the names of father/husband/guardians of individuals, as though all adult Indians are in fact less than adults and need someone else to be responsible for their behaviour.

Adult Indians also demonstrate this "non-adult" assumption. If trains run late, daily passengers squat on the track and disrupt the entire railway operation in that sector. It stems from some kind of a covert assumption that the powers that be actually do not care for the people and therefore the only way they can be made to take steps to remove people's difficulty is by disrupting administration. This kind of behaviour also seems to be based on the unconscious assumption that there is an unexamined belief in an alien power governing the country even though they are people's representatives.

It is an open question as to what extent these people's representatives too – those who get elected to the parliament and the various state legislatures – unconsciously act as though they are aliens with power to rule over the country's population.

The impact of the colonial hangover manifested in terms of adolescent behaviour, and its opposite, treating adult Indians as adolescents by the powers that be can be seen in everyday life if one cares to reflect upon one's experience.

There are regular media reports of people attacking employees and breaking up furniture and machinery in electricity supplying sub-stations if there is a failure in electricity supply during an important cricket match that was being shown live on a television channel, attacking hospital employees and breaking up equipment if one believes that a patient has died due to negligence, and so on. The assumption that there is no legal recourse to set them right could also be true to a great extent, perhaps because the powers that be also feel that they are ruling aliens. And the prize here goes to Kolkata, earlier known as Calcutta. While all over West Bengal, college students (who are all adults) and sometimes their parents (who pay the fees) run amok destroying college properties to force the authorities to dilute the standard of intellectual content. In Kolkata, in addition, educated and moneyed people who drive four wheelers and two wheelers not only break traffic laws, but start a serenade of motor horns every time there is a traffic jam and also as soon as the traffic light turns green. It is a phenomenon of not even adolescent behaviour, but infantile behaviour. It is the greedy child's show of temper tantrums at its inability to postpone gratification.

Like the greedy child interprets adult behaviour as "getting away with their gratification", in Kolkata, the adults who actually get away with their gratification are the police, the army, many government officers, and the Kolkata Municipal Corporation. Their vehicles routinely break traffic laws with impunity by driving up one-way streets in the wrong direction, parking vehicles in no-parking areas, and sometimes even driving by the wrong side of the road to avoid a traffic jam when they have a red light attached to the car – a phenomenon that one Kolkata based newspaper had termed "the red menace". The city police also has its own interpretation of traffic rules that differs vastly from what is taught by motor-driving instructors. Bicycles, rickshaws, push carts, and sometimes motorized two wheelers are not considered as vehicles and therefore allowed to travel in the wrong direction in one-way streets. Further, the practice of cutting corner while turning right into another street and the belief that by giving the right turn signal the motorcar driver receives the authority to turn right even if it means that the oncoming traffic has to stop, are accepted by traffic police as normal behaviour.

Then, Kolkata, the capital of West Bengal, perhaps leads the rest of the state in taking law unto oneself, whoever has the contextual power, as the legacy of its colonial history where those who had contextual power, i.e., the colonial masters, whether they were Afghan, Mughal, or British rulers, who had scant regard for laws of their own making when it came to their self-interest.

During the British colonial era, a great Maharashtrian leader of the freedom fight had said "what Bengal thinks today, India thinks tomorrow" to highlight Bengal's contribution to the struggle for freedom. If that saying becomes true once again, the Republic of India may be in for a lot of rough weather. One has to read one's daily newspaper thoroughly in order to look at some of the evidence of that scattered all over India.

9 Bureaucracy in Modern India

A Cultural Analysis

P.K. Misra

Bureaucracy has been defined as a rational system. It establishes a relation between legally established authorities and their subordinate officials which is characterized by defined rights, duties, responsibilities prescribed in written regulations. Relations between different sections of officials are ordered systematically. Appointments and promotions are based on contractual agreements and are accordingly regulated. It is supposed to be impartial as well as efficient. Bureaucracy has sufficient checks and balances to prevent growth of unbridled power and illegal gratification. This is the ideal version of bureaucracy. The reality is far different. The term bureaucracy is generally used in pejorative sense. It is used to denote torturous procedures, narrow outlook, and autocratic ways of the government officials. Bureaucracy everywhere has grown to monstrous size. Parkinson formulated several laws, one of which says that bureaucracy grows in inverse proportion to the work to be done (1957). Parkinson has shown how the officials work for each other, without enhancing the output. Rise in civil service has much less to do with the growth in the volume of work. There is much duplication of work and ultimately there is much concentration of power in few hands. The common man suffers most on account of bureaucracy. They are made to run from pillar to post.

In spite of the fact that there is so much negative impression about bureaucracy, it has stayed and grown everywhere. One of the things which the modern world has in common across countries is bureaucracy. But each country has grown in a particular historical, social, and cultural context. Therefore, everywhere it has its own style, nuances, emphasis, targets, strengths, weaknesses, and so on, in brief its culture.

Indian context

Modern bureaucracy in India is a legacy of the British Raj, but the context in which it operates after India became independent has considerably changed. It will be worth mentioning some of them for the purpose of my discussion here. India opted for democracy. It has a constitution, one of the guiding principles of which is equality before law. The country opted to be a welfare state. Development is one of the important programmes of the state, of which the welfare and development of the backward classes is the specific target. To facilitate this objective, the constitution has made special provision which is popularly known as positive discrimination. Certain communities were identified for welfare programmes as well as providing reservations for the members of these communities in elected bodies, government offices, and educational institution. Previously, there were communities under two categories namely scheduled tribes and scheduled castes. But later on it was found that besides them there were several other groups who deserved to be included under the scheme of positive discrimination and hence new categories came into existence such as Backward Classes, Other Backward Classes, Economically Backward Classes, Most Backward Classes, and so on. How the new categories came into existence is a topic by itself, suffice here to say that the process of inclusion means erecting some boundaries which are bound to be found inadequate in course of time for a variety of reasons. Also, inclusion and exclusion are dialectically related. However, one of the indices that are used to demonstrate the backwardness of the communities is their percentage of representation in the bureaucracy. It may be stated that each state in the country has taken up the task of updating the list of backward classes in their states. For this purpose, Backward Class Commissions are set up. Even before the ink dries up in the recommendations of the commissions some objections on their recommendations, and demand for inclusion or exclusion of some communities are made. The intervention of judiciary is also sought. As a result of all these, a case for a fresh commission is made. Thus, we see the specter of commissions after commissions to determine the backwardness of communities a seemingly unending debate. It has been argued that for good and 'understanding administration' adequate representation of the communities is essential. The counter point is that instead of representation, competent and meritorious personnel are required in services. I will not get into this issue here as much has been written on it.

There are some other institutions also which directly impinge upon bureaucracy in modern India. They are elected representatives of various legislative bodies. The elected representatives are the voices of the people. Not only they have to nurse their constituencies, listen to the grievances of the people, and enact suitable laws, but also draw the attention of the governments wherever they find there are lapses and lacunae in the implementation of policies and administration of the government. It has been a considered view that too much power is concentrated in the few instruments of government which is neither conducive for efficient and responsible administration nor for development. Sir Mirza Ismail stoutly opposed the dissolution of Representative Assembly in old Mysore. He wrote, "by killing all local loyalty and initiative you do not strengthen but weaken the country" (1954: 23). The decentralization of power has been an issue in India right from 1955. The thinking is that more and more people at grass-root level should be made responsible to look after their own affairs. On paper through legislation for Panchayat Raj, it appears massive decentralization of power has been accomplished but the ground realities are different which need not detain us here. The other institutions worth mentioning are judiciary and media. Their role and effectiveness in administration have tremendously increased since the country became independent.

Objective

The above roughly is the context in which bureaucracy in India has to operate. Since the topic of bureaucracy is vast and rich, I propose to limit my talk to two aspects, namely, hierarchy and values. The relationship between structure of the society (hierarchy is one of its aspects) and values is complex but suffice here to say it is interdependent. Hierarchy in bureaucracy is designed so that it works smoothly and efficiently. It has clear cut hierarchical division of authority, power responsibilities, and duties. In the western society, it appears that there is a mismatch between the ideology of equality, competitiveness, sense of freedom, and adventure, accumulation of wealth, etc., and the hierarchy in bureaucracy. In India, where the western concept of bureaucracy was planted, had an ideology of inequality and a complex traditional hierarchical structure which in spite of many changes in the larger system continue in modern India though in a bit abbreviated form. My hypothesis is that there is a close fit between the two in the Indian situation. Dumont writes, "modern western society presupposes equality to the point that delicate situations are likely to arise where subordination is necessary. India on the contrary emphasizes inequality to the point that situation tending to equality are unstable and conflict is called for to solve them by establishment of gradation" (1970: 14).

Hierarchy in bureaucracy

Bureaucracy has hierarchy from peon to the top men around ministers – the decision-makers. Those of us who have seen "Yes Minister" would know that there are innumerable ways by which the bureaucrats insure that ultimately the defacto power of decision-making does not slip away from them. Moreover, a minister has a variety of roles and in each, he is expected to be an expert. He as a head of the ministry is expected to be the authority on the subject. He has to pilot the bills concerning his ministry and is answerable to the house, media, and public at large. For even the most sincere and hard working minister to perform all these roles as an expert and authority is an impossibility. In reality, the minister becomes a spokesman and defender of bureaucracy. During the course of his duties, a minister has to sign a large number of papers.[1]

The simplified version of the bureaucracy in India, as in the case of the caste system, is divided into four classes – Chaturvarna. But in between these classes, there are innumerable sub classes identified by their scales of pay,[2] authority, power, perks, and various symbols. There is a continuous intense jostling among them to rise above their ranks except for class IV employees[3] where jostling is minimal. This jostling results in upward mobility for some individuals, but jostling itself is a cause of much action or inaction in the government organizations.

Class IV

Safaiwalla, peons of various descriptions, messengers and daftari are some of the ranks among the class IV employees. But safaiwallah cannot rise to any other rank. On the whole, they are the untouchables of the Indian bureaucracy. They are not only at the bottom of the rank order but also continue to, come from a particular cluster of castes of the Indian society.[4] This class of employees is uniformed. The uniform of this class of employees who are around VIPs are a bit more distinctive and may have to wear some kind of cross belts and head gears. In non-VIP sector, wearing of uniform is considered below dignity. Hence most of the class IV employees do take their uniform and also allowances for their washing but rarely wear them. Officially, the duties of the class IV employees are manual work such as carrying files and papers from one desk to another, act as messengers and so on, but everywhere they do much more than that. Generally, they are not provided any "desk" or "chair", at best they may be provided with some stools or discarded chairs outside the chambers of the bosses. Provision of a peon for an officer adds to the status of the officer.

In taluk, district and divisional offices of the state governments in India, peons wield considerable powers. It is well known that they would be helpful in making papers move from one desk to another, can help visitors to meet the officers, and can even by subtle moves influence the decisions.[5] However, before the peons get into offices to work, they may have to work several years as household servants in the "bungalows" of the officers. It is here they learn much about the officers, the kind of work they do and begin to enlarge their area of influence in the office of the officers. They are also responsible for creating folklore about their bosses. Of course not all are so fortunate to come out of "household" duties but those who do begin to command respect among the fellow peons and "babus". Though they are untrained, they are the nearest observers of their "masters" and those who interact with them. The world of the "peon" is potentially a rich area of research which has remained unfortunately uncharted. The world of the "peon" would give a subaltern view of bureaucracy, how it makes decision, the gossips it indulges in, and so on.[6] One of the interesting things would be to examine what the peons think of what earth-shaking things go on in offices that their bosses are so well provided, while they have to make innumerable errands to bring coffee, snacks, cook food, etc. as part of the office duty.

Class III

Howsoever interesting may be the subject of class IV, I must quickly move on to the next category, in the official parlance they are called class III employees but popularly are known as "babus". They form the bulk of the government employees. Of course in any office they are more in numbers than required but unlike the peons they have desks and chairs. Each one of them is supposed to have defined tasks to do but there is no finality about it. The tasks keep on dividing, subdividing, and evolving in the history of any office but they are interlocked and stratified.[7] Hence, it is important that the rank of the babu in terms of his/her seniority is kept in view before the allotment of task to him/her. This is the class which not only on account of its bulk but also the way it does the office work determines the pace of work. The top men may cry hoarse but the file has to be put up. The file cannot be put up unless it has passed through all "gates" in highly ritualized "downward" and "upward" movements. The chain cannot be broken because each "stop over" has a certain responsibility and each has to put its observations howsoever repetitive they may be. In the process, the file accumulates weight and sets its precedent. Howsoever critical one may be of torturous and long procedures of file movement, it should not be forgotten that file at any time can be examined by innumerable senior officers, vigilance officers, task forces, judiciary, and now the Right to Information Act has added a new dimension to the process of noting in the files. Therefore, the file work should be "well done" howsoever far it may be from the task at hand. Whosoever gains even a little bit of experience in bureaucracy, begins to rely on the first mantra that self-protection is the primary rule and second is to pass on the buck of decision-making to the next officer. However, the movement of the file can be hastened. Any hastening process requires lubrication or influence or both. This has been called as corruption but it is too simplistic. It requires a detailed societal analysis. In Uttar Pradesh, one of the most forward states in India in coining slogans, the practice of lubrication is called as "suvidha shulk" (fees for facilitation).

Class II and I

The class II and I are the official classes. Most of them are recruited through state and all India competitions; of course some can slip through into these classes through promotion or direct appointment depending on their ability to influence. Like the other classes, there are sub-classes here too. While the officer class is important by itself, the relative importance depends upon the post, power, perks, and various other factors.

This class has created a variety of symbols of its status such as strength of personal staff, quality and size of office furniture, residential accommodation, chauffeur-driven white ambassador cars,[8] phones, modern gadgets, personal toilets, etc. In an episode, Upamanu Chatterjee tells that a candidate on being asked in an interview by the members of the selection board as to why he wanted to join the civil services said "because within civil services, one is likely to have a peon, a personal assistant, and an ambassador car as buffers between one's good self and the rest of the government" (2000: 170). These symbols, as if the end in itself can cause much rivalry, jostling, and competition among the officers' class. There would have been finality about the jostling if the symbols of status were static. Every now and then new symbols are added to the extant list such as mobile phones, a certain category of "security", and so on. I can do no better than again quote Upamanu Chatterjee, who said that "district administration in India is largely a British creation, like the railways and the English language, another complex and unwieldy bequest of the Raj. But Indianization (of a method of administration, or of a language) is integral to the Indian story. Before 1947, the collector was almost inaccessible to the people; now he keeps open house, primarily because he does a different, more difficult job. He is as human as flexible, but now others can tell him so, even though he still exhibits the old accoutrements (but now Indianized) of importance – the flashing of orange light on the roof of the car, the passes for the first row at sitar recital, which will not start until he arrives and for which he will not arrive until he has ensured by telephone that everyone else who has been invited has arrived first. In Madna, as in all of India, one's importance as an official could be gauged by how long one could keep a concert (to which one was invited) waiting" (1989: 10).

As most of the officers eventually tend to get hold of the symbols of status, the concern for "higher rank" is not allowed to die. When it is found that everything is almost the same, the relative rank is established on the basis of the batch of recruitment, a specific appointment, and so on. The consideration of hierarchy in civil service is so pervasive that it gets reflected in almost all aspects, as if that is the only goal of bureaucracy.

After the country became independent, the central government at Delhi constructed a large number of residential accommodations for its employees in different colonies. These colonies not only made clear cut distinction in the layout, area of accommodation, material used, etc. but were named as Shannagar, Mannagar, Vinaynagar, and Sevanagar. Later, somebody pointed out that it was too blatant and therefore the names were changed but not the exclusivity and style of each colony.

Bureaucracy and representativeness

India faces two serious challenges, one is mass poverty and the other is socially discriminated population. In order to face these challenges, various welfare programmes, specifically to reach out to the last person in the row, were adopted. For the second, as already stated, the policy of positive discrimination was launched. It was also argued that once the representatives of the erstwhile socially discriminated population reached to the various positions in legislature, bureaucracy, etc., they would be in a better position not only to highlight the problems faced by such populations but also protect them. This in theory is certainly a sound policy but apart from other spheres it has caused some unintended consequences on bureaucracy. In order to make a case for not finding enough representation in bureaucracy, many caste organizations have been putting their cases before the various authorities. In order to have some objective data, various SC, ST, and backward-class communities began to not only count the numbers of each caste in the total bureaucracy but also in its different ranks. In other words, whether we like it or not, the policy of positive discrimination directly linked bureaucracy with the factor of caste. As we all know caste is a highly complex system, Varna is its simplified version. There are innumerable castes and sub-castes.[9] New castes keep on emerging owing to a variety of reasons such as a caste may become too large or some of its members may migrate to a new locales, or some of its members may violate some caste norms and thrown out of caste or give up their traditional occupation or a section of them may begin to sanskritize their customs and practices etc.

If it is conceded that bureaucracy which is the instrument of governance has to be representative of its people, the problem of the problem is how to represent the complexity of caste in it. The problem gets compounded because new castes keep on emerging. Further, the positions in bureaucracy have different levels of significance and therefore how does one resolve the problem of giving position of same significance to all the castes. Impossible as it may appear, but the same impossibility has become *raison de etre* for political activities. Many consequences can be deduced from trying to find a solution to the unresolveable problem such as enlargement of bureaucracy, keeping the populations in perpetual stage of agitation, incompetent people rising to the highly technical position, corruption, transfer becoming an industry, etc.

Here, I am basically concerned with governance and representativeness. If the bureaucracy is not representative, objection can be raised about its lack of understanding in its governance. If it is representative, still objections can be raised for its too much understanding. The apathy shown by bureaucracy in dealing with the recent communal riots in Gujarat illustrates the danger of representativeness.

Way back in 1970s, I was participating in a marriage of my friend's sister. At that time because of scarcity of food, the Uttar Pradesh government had promulgated a rule that no one would feed more than 25 people in a marriage party. But almost everybody violated this rule rather openly. However, some took precautions to circumvent the promulgation by printing on the marriage invitation card phrases such as "please bring your own ration" or "no cereals will be served" etc. Besides, the invitees were fed in batches of 25. In that particular marriage, I found a police contingent along with a head constable at the marriage venue. (There was an order from the state government that during the marriage season, the police should inspect the various marriage booths so that the orders of the state are not violated. Of course, always they were the first to be fed hurriedly so that relieved of partaking of food they may enjoy inspecting other marriage venues.) As the police officer finished his meals, I walked near him and asked about the restrictions. He almost asked me to shut up and without any hesitation said that how could he cause any hurdle in the marriage of a "kanya". He too had grown-up daughters.

Implementation of anti-dowry law is not different. It is well known that in a large number of communities, dowry is taken and given. In the eyes of the law, dowry givers, takers, as well as those who are witnesses to the act are criminals. In Uttar Pradesh, they have found as easy solution to the problem since no one wants to be present at the time of receiving or giving dowry though they may actively participate in its negotiation, the ceremony takes place in a Hanuman temple – the bachelor god stands as a witness. In Kerala, I am told among certain Christian groups, it is the church priests who negotiate the dowry amount. The church has a vested interest in these negotiations as it gets a certain percentage from the dowry. How a representative bureaucracy or otherwise can take action against the deep-rooted tradition and values associated with them.

Values

The dominant values in the Indian society are to maintain inequality, curb competitiveness, curtail individual freedom and initiative, instill the idea of satisfaction, instill fear of unknown, discourage adventure, and discourage the spirit of accumulation of wealth. There has been a close fit between these values and the traditional structure of the Indian society. It is not possible for me to discuss in detail the family and caste, the two important aspects of the Indian society here to demonstrate how closely they fit with the values of the society. Suffice here to say that the relationship between the different members of the family is based on the concept of inequality, curbing individual freedom, showing deference to elders and those who stand in higher rank, etc. The members in the family are expected to observe restraint in their behaviour. Achievements are not to be exhibited but to be considered as "gift of god". Similarly, inter-caste and intra-caste relations are based on the concept of inequality, non-competitiveness, and recognition of the social and cultural boundaries and symbols. No doubt that some significant changes have taken place in all aspects of the Indian society and there are vast individual differences, but it appears that the basic structure and the values have remained intact, of course with some changes.

If this is correct then it can be argued that there is no friction between the structure of the Indian society and bureaucracy in modern India. I have earlier shown that bureaucracy is not only stratified in structure but its concern for rank order is pervasive. It has created innumerable symbols of that and has attached meaning to them.

It is possible to argue that those have become the real objectives. There is no aberration between modern bureaucracy and the Indian society. In a way, I am prepared to argue that it has strengthened its traditional structure. It would sadden those who have been cherishing an illusion that bureaucracy in India would become an instrument of radical change.

I would like to conclude this by quoting a passage from Sir Mirza Ismail: "If the Asian countries are to be real world-states, and themselves to attain prosperity and happiness, they must give concentrated attention to their domestic policies. All of us are suffering from sickness of the body politic. That sickness is largely due to the personal corruption of men in power. Here, no half measures can be tolerated. We must have integrity in all leaders and throughout the services.

Without either remorse or fear, we must get rid of every corrupt person" (1954: 165).

The above observations of Sir Mirza Ismail made more than half century ago as an experienced administrator need no elaboration in modern India.

Notes

1. A minister told me that in a course of day he signs hundreds of papers which he does so mechanically that he may even sign his own death warrant which perhaps he would come to know only at the time of its execution.
2. I am told that there were more than 300 scales before Third Pay Commission. They were compacted and reduced to 35 in the Sixth Pay Commission.
3. These days four-fold divisions of the bureaucracy is called by different nomenclature such as ABCD, groups of employees.
4. Way back in 1970, my friend who was the director of a social science research institute under the government of Kerala pointed out that recently his institute had appointed a Namboodri young man as safaiwala. His name was sponsored by the employment exchange. He further told me that the boy himself did not clean the toilets, swept rooms, etc., he hired people of scavenger caste to do the job for him.
5. Several decades ago a friend of mine was posted as an IAS trainee officer in Mysore. He and his wife were fond of seeing films. In those days, perhaps that was the only source of entertainment for a middle-class family. He always sent his peons to get "tickets". He said that saves embarrassment from both the sides. Of course, no one bothers to find out what "cut" the peon derives in such short cuts.
6. I am reminded of a study on the life of an untouchable in a village in Orissa conducted by an anthropologist. His subject relates in first person how he was used by high-caste priests, landlords, and government officers to procure "untouchable" girls for them.
7. Parkinson pointed that official staff expands in inverse proportion to the work to be done. Officials create work for each other.
8. It had changed to Contessa car which has changed to more expensive cars now.
9. Anthropological Survey of India conducted a survey and found there were 6,748 communities in India (Singh, 1992).

References

Chatterjee, U. 1989. *English August: An Indian Story*. Calcutta: Rupa.

Chatterjee, U. 2000. *The Mamaries of the Welfare State*. New Delhi: Penguin.

Dumont, L. 1970. *Religion/Politics and History in India*. Paris: Mouton.

Ismail, Sir Mirza 1954. *My Public Life: Recollections and Reflections*. London: George Allen & Unwin.

Singh, K.S. 1992. *People of India: An Introduction*. Calcutta: Anthropological Survey of India.

Index